WASTE MANAGEMENT
IN THE OIL INDUSTRY

WASTE MANAGEMENT
IN THE OIL INDUSTRY

ALOYSIUS A. AGUWA, Ph.D.

iUniverse, Inc.
New York Lincoln Shanghai

Waste Management in the Oil Industry

Copyright © 2007 by Aloysius Aguwa

iUniverse books may be ordered through booksellers or by contacting:

iUniverse
2021 Pine Lake Road, Suite 100
Lincoln, NE 68512
www.iuniverse.com
1-800-Authors (1-800-288-4677)

ISBN: 978-0-595-41109-2 (pbk)
ISBN: 978-0-595-85847-7 (cloth)
ISBN: 978-0-595-85466-0 (ebk)

Printed in the United States of America

CONTENTS

LIST OF FIGURES

LIST OF TABLES

ACKNOWLEDGMENTS

During the preparation of this book, I was assisted by many professionals in the field of waste management. I acknowledge the assistance of Mrs. Melissa D. Cruickshank (geologist), Mr. Arinze Nwamba (chemical engineer), and Mr. Marquez Cruz (civil engineer) for their research and development efforts toward completing this book. Special thanks to Ms. Manasi S. Koushik (environmental engineer) for her assistance with the research, development, editing, and reviewing of this book. Many thanks to Christopher C. Ohanele (geologist), Mark Resch of Altech Environmental Services Inc., and John Pentilchuk of Conestoga-Rovers & Associates for providing constructive technical comments.

1.0 INTRODUCTION

While our quality of life has unquestionably been enhanced by, and indeed has come to depend on, the many products of the oil industry, it has at the same time become vulnerable to the pollution generated by the various processes involved. The pollutants generated by the industry are perhaps even more diverse than the products that are produced by the industry (e.g., crude oil, natural gas, refined products). The variety of toxic organic and inorganic wastes that are generated has the potential for degrading all environmental media—air, water, and land.

The primary goal of this book is to present practical information on the management of waste generated by oil exploration, production, refining, distribution, and storage. The presentation includes not only a state-of-the-art review of available management strategies, but also a review of how wastes are generated as a result of the various processes.

The exploration, recovery, and utilization of hydrocarbons is a complex, step-by-step process consisting of several stages of operation as shown in the simplified schematic in Figure 1.1. Simplified, these operations can be broken down into four general activities: exploration, production, refining, and distribution. Exploration and production (E&P) constitute "upstream" operations. In E&P operations, raw fuel sources (crude oil, natural gas) are sought out, possibly located, extracted from the subsurface, and transported (distributed) from the well to the flowstation and on to the export terminal or to the refinery for further processing. Refining and distribution (R&D) are "downstream" operations whereby refined and useful fuel products are manufactured and ultimately transported to the consumer. Several additional operations are associated with each of these four activities.

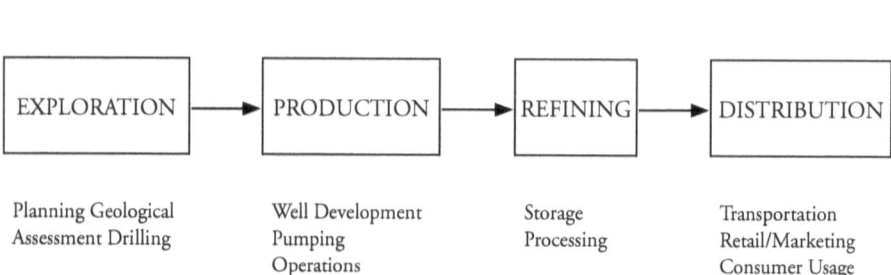

Figure 1.1   Operations Associated with Exploration, Recovery, and Utilization of Oil

Petroleum exploration takes place in two broad terrains: offshore and onshore. Offshore exploration is concentrated in shallow or deep marine environments. Onshore exploration occurs on continental landmasses, either on dry land or in swamps. Exploration techniques are varied but can be generally characterized as non-invasive and invasive. Almost all invasive exploration techniques consist of test hole drilling, in which the prospective area is physically penetrated with a drilling assembly. Non-invasive exploration methods use techniques that are administered on or perceived from the surface of a prospective area.

If exploration efforts prove the presence of hydrocarbon accumulations, those accumulations are then evaluated for their commerciality, taking into consideration such factors as the volume of recoverable reserves, production costs, prevailing prices of crude oil and natural gas, and availability of a market in the case of natural gas.

Production of oil from a well is the next procedure once the oil accumulation is found to be in commercial quantities. The step from a successful well to production could take days or even years. As soon as drilling is completed and all necessary data are acquired, a wellhead is installed, and the drill rig can be removed. The wellhead is necessary for the safety of the well, as it checks the enormous pressures in the well and so controls the flow of hydrocarbons from the well. Once the wellhead is installed, the oil will be transported off-site whenever the decision is made to start production. Pumping equipment may or may not be required to bring the oil to the surface. The crude oil may initially be taken to a flowstation, where crude is stabilized and the fluid components (oil, gas, and water) are separated.

Pipelines are the modes of transportation used to transfer oil from the well site or the flowstation to an export terminal or refinery where it may be stored in tanks (tank farms). In some cases the oil may be transported by barges along navigable inland waterways. Export crude is loaded into tankers at the terminal.

The purpose of a refinery is to refine crude oil into marketable products such as gasoline, kerosene, fuel oil, and so forth. A storage facility or tank farm may be used to control the flow of oil through the refinery. The operation of a refinery consists of many processing steps. Atmospheric and vacuum distillation of the crude oil begins the process. The processes of catalytic reforming, fluid catalytic cracking, alkylation, delayed coking, extraction of aromatics, thermal hydrode-alkylation, hydrogen treating, and sulfur recovery will further refine the distilled oil.

The final stage is distribution, in which refined products are ultimately shipped to consumers through various modes of transportation. These modes include trucks, pipelines, ships, and railroads. The mode of transportation selected depends on the type and quantity of product and the rate of delivery required by the consumer. Because upstream and downstream activities entail uniquely different operations, varying wastes and environmental issues are associated with each operation.

2.0 OIL INDUSTRY FUNDAMENTALS

2.1 Exploration

The quest for oil has driven geologists to explore the earth for over a century.[1] From the earliest days of the wildcatters who used divining rods and rudimentary knowledge of earth sciences, to the modern era of technology, exploration has evolved into a precise undertaking of immense proportions in terms of both resources and financial investment. Waste management is a critical element of oil exploration, with respect to efficiently using resources and preventing potential environmental damage that may result from intense investigative programs.

History of Exploration

The earliest geological exploration efforts mainly focused on mapping and assessment of surface geological features.[4] Oil seepages, the most readily identified surface features, were most sought-after, as they were the most basic indicators of oil. In fact, the monumental discoveries that triggered the beginning of the petroleum age, such as in northwestern Pennsylvania, USA, were the result of exploratory drilling in regions where local populations had long recognized the usefulness of these surface petroleum seeps for waterproofing and for illumination purposes.[2, 3] The first major oil fields were large, economically rewarding, and easily detected at the earth's surface, thus ensuring successful exploration. However, as competition increased and resources were depleted, the need grew to expand exploration. As early as 1920, American oil companies driven by competition within the industry began using geophysics in the hunt for oil.[2] Similar records exist for geophysics use in India. However, widespread use of geophysics in oil exploration did not begin until the post–World War II era.

Oil Formation

Whether on land or at sea, the goal of exploration is to find geologic formations that have the geometric or stratigraphic configuration favorable to the trapping of oil and gas. Further investigation in the exploration process seeks to determine whether hydrocarbons are likely to be present and the chances of those hydrocarbons being oil or gas. Geologists now employ a number of predictive tools to minimize risk and improve the chances of success. Eventually an exploration well is drilled to confirm the presence of hydrocarbons.

The most favorable targets for exploration are sedimentary basins—vast regions of layered sedimentary rock deposited over extended periods of geologic time, trapping organic materials in quiescent pools. Continual deposition of sediment within the basin creates an oxygen-deficient environment, allowing microorganisms to further decay the trapped organic materials into carbon-rich deposits. The process of deposition, compaction, and subsidence means that over a long period of time a very thick pile of sediments, sometimes tens of thousands of feet thick, is formed. In a clastic setting, over time, increasing heat and pressure begins to transform these deposits, now deeply buried, into carbon-rich shale, the source rock for oil and gas. Further deposition increases heat and pressure on the source rock, causing it to reach maturation and to expel crude oil and natural gas into the surrounding porous sedimentary rock. Oil and natural gas tend to migrate to the most porous layers of sediment, called the reservoir rock. The most common reservoir rocks are sandstones and limestones. For an oil deposit to be of economical size, an impermeable layer must encompass the reservoir rock, forming an oil trap. Figure 2.1 is an illustration of this process.

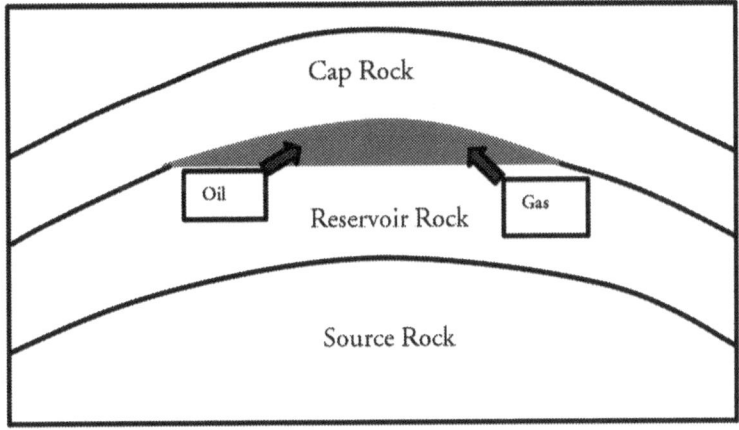

Figure 2.1   Oil & Gas Formation

2.1.1 Geophysical Exploration

Geophysical exploration for oil involves using technology to assess the physical characteristics of the subsurface by measuring physical properties of the earth's surface.[5, 6] Although these methods do not directly indicate the presence of oil and gas deposits, they do provide geological clues for geologists to pinpoint drilling targets in sedimentary basins that are most likely to contain hydrocarbons. The judicious use of geophysical exploration programs is crucial for oil companies to be profitable and to expand their resource base.

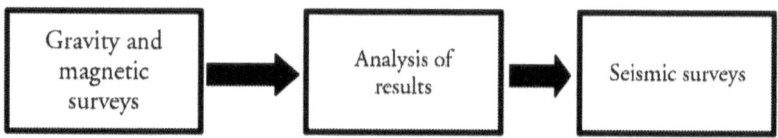

Figure 2.2   Geophysical Exploration Hierarchy

Geophysical techniques are typically executed in a hierarchical fashion as shown in Figure 2.2. In the very early stages of exploration in a new sedimentary basin, gravity and magnetic surveys are performed first to delineate large-scale geological structures, such as the perimeter of a sedimentary basin of interest.[7] Gravity surveys measure deviations in the earth's gravitational field that are a result of the density differences in geological structures. Magnetic surveys detect the differences in the earth's magnetic field that are caused by differing magnetic properties of geologic formations.

Both surveys are performed with highly sensitive measuring instruments. Gravity surveys can be executed from land or sea, while magnetic surveys are performed from the air, either with satellites or airplanes. Due to the non-invasive nature of these surveys, waste generation is at a minimum.

Upon completion of gravity and magnetic surveys, the data are analyzed and interpreted to define the boundaries of the sedimentary basin of interest. These data are used to lay a framework for subsequent, and much more costly and involved, seismic surveys. In some cases, when exploration is being performed in areas of known oilfields, gravity and magnetic maps may already exist and may be available for purchase.[6]

Careful analysis of data obtained through gravity and magnetic surveys allows geologists and geophysicists to plan for seismic surveys. Seismic surveys are performed both onshore and offshore. The basic premise of the seismic survey is to generate seismic waves with, for instance, explosions, and to use highly sensitive instruments to measure the waves' paths as they travel through the subsurface.

Seismic surveys begin by arranging parallel lines of highly sensitive instruments (geophones on land, hydrophones on water) to detect and record wave patterns through the subsurface.[7] For land-based surveys, a ground crew must remove all impediments (vegetation, etc.) in order to rest the line of geophones on the surface. For sea-based surveys, floating lines of hydrophones are carefully placed on the water's surface. Historically, dynamite charges were detonated both on land and at sea to produce seismic waves. However, in recognition of the safety hazards and environmental impacts that result from dynamite usage, technology has evolved to replace the use of dynamite to generate seismic waves with the vibroseis method and compressed-air guns. On land the vibroseis method is used, in which large trucks equipped with heavy plates ("thumper trucks") propel energy into the earth to create seismic waves.[8] At sea, compressed-air guns are used as an energy source, greatly decreasing the risk to aquatic life.

Upon completion of the seismic survey, the data is analyzed with highly specialized computer programs to generate detailed geologic representations in the form of cross-sections. Each line of instruments used in the seismic survey produces a two-dimensional (2D) picture of the subsurface geology. When closely spaced grids of multiple survey lines are used, a three-dimensional (3D) model of the subsurface is produced, illustrating such features as oil traps and reservoir rock. Recent technological advances have incorporated the variable of time into the seismic survey by repeatedly performing 3D surveys along the same lines over periods of months or years, producing four-dimensional (4D) results. A 4D survey provides an illustration of geological changes over time, such as the migration of oil.[6, 7, 8]

Although great care is taken in executing a seismic survey, some environmental damage is inevitable. For instance, proper placement of the geophones during a land survey most often requires devegetation of the investigation area.[8] Left untreated, this can lead to soil erosion and deforestation of affected areas, as well as the interference with and possible destruction of animal habitats.

2.1.2 Exploratory Drilling

With basic definitions from gravity and magnetic surveys, and detailed geologic information from seismic surveys, the exploration program may be further expanded to include drilling exploratory wells.[7] Because this is the most expensive part of the exploration program, drilling is not performed until after working through the exploration hierarchy and analyzing all of the data obtained from field observations and geophysical exploration programs.

The drilling location is planned for the region most likely to yield oil. The first well is commonly referred to as a "wildcat," in reference to the earliest days of oil exploration, when wells were drilled in unknown regions, planned only with the hope of discovering an oil field.[2] If the well does not produce oil, it is referred to as a "dry hole." A dry hole is not a complete loss to the oil company, as it can still yield important geological data about the region, which can later be incorporated into future exploration programs.[1, 8] However, if the well does produce oil, it is called a "discovery well" and will be further evaluated for commerciality of the discovery before proceeding to production.

2.1.3   Drilling Operations

Drilling History

Drilling technology was invented in ancient China to obtain salt brine from deep subterranean deposits.[1] The Chinese used a heavy drill bit connected to a wooden platform with a rope. The drill bit was driven into the ground by teams of men jumping on the wooden platform. This method was a direct ancestor of cable tool drilling, used successfully for the first oil-producing fields in North America throughout the nineteenth and early twentieth centuries.[2] Cable tool drilling incorporated a heavier drill bit made of steel, a wire rope, and a heavy drill stem connected to a fulcrum and supported by a wooden frame structure. The weight of the drill stem propelled the bit into the earth to produce boreholes. Water was added to the advancing borehole to provide lubrication for the drill bit. During production with the cable tool method, drilling would stop every few feet for drillers to manually remove rock cuttings. This method also pioneered the use of drill casings and hollow tubes, which are used to line the hole to prevent cave-ins.[1]

Cable tool drilling was dangerous due to the occurrence of blowouts, in which excess pressure within the borehole would cause oil and rock cuttings to spew out in highly pressurized flows. Blowouts resulted in serious injuries or fatalities to the drilling crew and massive, virtually unstoppable losses of oil.[1, 2] Cable tool drilling was also limited to shallow depths and involved relatively slow drilling progress.

The introduction of rotary drilling, which used a rotating drill bit to cut through the subsurface, revolutionized drilling practices and the oil industry as a whole.[2] By mechanizing the drilling process, greater depths could be reached with greater speed, which is important for production and resource extraction. Furthermore, rotary drilling addressed safety concerns and waste prevention from blowouts by incorporating pressure-regulating technology involving the use of drilling fluids and the installation of a blowout prevention device (BOP) on the rig floor.[2, 3] In stark contrast to these advantages, rotary drilling also increased the amount of drilling-related waste, resulting in the need for strict and carefully planned drilling-waste management programs.

Rotary drilling gained the attention of the oil industry after it was successfully used in 1901 to drill the first wells at the Spindletop oilfield in Texas.[4] Initial exploratory attempts using cable tool rigs failed miserably due to the very soft upper sand and clay of the Spindletop formation, which would collapse and impede drilling progress. The rotary drilling rig, a much more durable machine capable of reaching greater depths with continuous drilling, was able to advance through the troublesome sand and clay layers. Part of the rotary drilling success was due to the introduction of drilling fluid, a solution created by drillers at Spindletop who noticed that the thick mud formed during drilling provided excellent support for borehole walls until the casing could be installed.

Rotary drilling is the most widely used drilling technique in the oil industry.[1, 2, 8] However, in smaller oil fields with shallower deposits, such as those in the Appalachian region of the United States, a modified version of the cable tool drill rig is still used.[1]

Rotary Drilling Components

The rotary system is named for the method of advancing the drill hole, which occurs by mechanical rotation of a metal bit through the subsurface. A rotary drilling rig has three main components: a drilling system, a mechanical system, and a fluid circulation system as shown in Figure 2.3.

Drilling System

The main component of the drilling system is the drill bit, which consists of either sharp metal teeth or industrial-grade diamonds to cut through the subsurface geology. Power to advance the drill bit is conveyed from the mechanical system to the drill string by the swivel, a rotating bearing. The drill string consists of a strong cable reinforced by thick sleeves of pipe called drill collars. As the borehole is advanced, 30-foot lengths of drill pipe are added to the drill string to increase the depth. The drill string is connected to the kelly, a stationary hexagonal drill pipe that joins to the rotary platform through the kelly bushing, a hexagonal device that contains a series of rollers to establish nearly frictionless advancement of the drill string. The kelly bushing is connected to the rotary table, a platform that is mechanically rotated to provide the driving force to advance the drill bit through the subsurface. Figure 2.3 depicts a typical drilling rig.

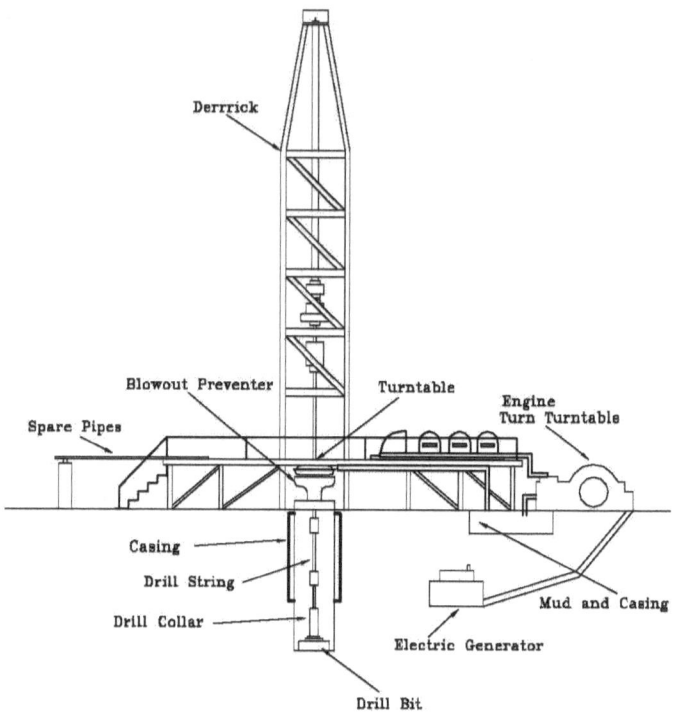

Derrrick

Blowout Preventer   Turntable

Spare Pipes   Engine
                                           Turn Turntable

Casing

Drill String

Drill Collar   Mud and Casing

                              Electric Generator

Drill Bit

Figure 2.3   Anatomy of an Oil Rig

Mechanical System

The mechanical system consists of hoisting equipment that is used to add or remove lengths of drill pipe and to access the drill string for maintenance or drill bit replacement. [10] The main component of the mechanical system is the draw works, a mechanized winch that is connected to the crown block, a traveling block-and-tackle pulley system located in the upper portion of the derrick. The draw works is connected to the crown block through the block line, a steel cable attached to a large hook. The derrick is a crucial part of the mechanical system, as it houses the hoisting infrastructure and provides structural support for additional lengths of drill pipe.

Diesel-powered engines provide power to the drill rig for rotation as well as to operate electric generators that drive the mechanical system.

Fluid Circulation System

The rotary drilling process introduced the usage of drilling fluids (also called drilling mud) to cool the drill bit as it advances through rock and to flush rock cuttings from the borehole to the surface.[1, 3] Although it was a great advance in drilling technology, the introduction of drilling fluids also increased the amount of waste generated during drilling. However, in an effort to reduce waste a sophisticated recycling system was employed in which fluids that had been run through the system were continuously reused.[10] This ingenious method served as both a cost cutting measure for the drilling company as well as a waste reduction strategy.

In addition to the use of drilling fluids, all rotary drill rigs are equipped with a device called a blowout preventer (BOP), a system of pressure valves located on the drilling floor.[1] The blowout preventer is important as a safety feature and as a waste-prevention device, because it helps to contain high-pressure drilling operations and prevents oil from gushing to the surface.

Drilling fluids are prepared on-site in the mud-mixing hopper by fluid technicians, and are piped to mud pits adjacent to the drill rig.[6] Fluid circulation begins when the mud pump pushes fluid through the hose that passes through the drill pipe and feeds into the drill string. Fluids enter the borehole through openings in the drill bit and are circulated back to the surface through the annulus, the void surrounding the drill pipe. At the surface, used fluids are passed over a series of moving screens called shale shakers, which separate rock cuttings from the fluid. A moving platform called the shale slide then deposits rock cuttings in lined reserve pits located near the drill rig. This process is schematically shown in Figure 2.4.

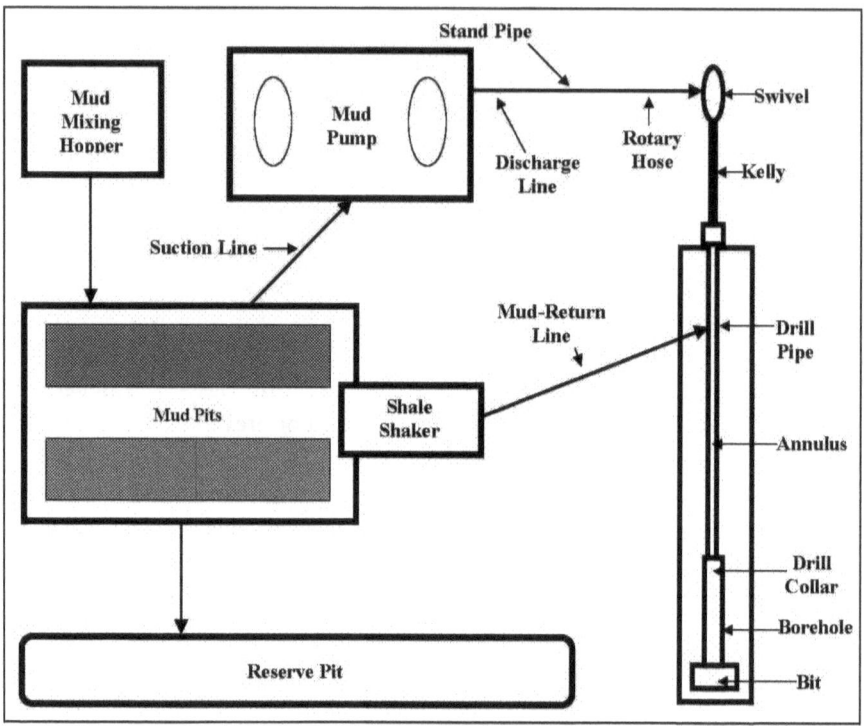

Figure 2.4   Fluid Circulation System Schematic

An integral part of fluid circulation systems is the installation of the casing, a large-diameter metal or concrete pipe that lines the borehole. By containing the drilling fluids in a casing, fluid circulation is improved, and fluid loss into the surrounding geologic formation is avoided.[1, 6] Preventing fluids from entering the subsurface is a critical step in pollution prevention and waste control. Additionally, the casing preserves the chemical integrity of drilling fluids by preventing groundwater from entering the borehole.[6, 8] The installation of the casing into the borehole is also important to drilling production, as it prevents the borehole from collapsing.

Drilling Fluids

Drilling fluids are used to flush rock cuttings from the borehole and to regulate fluid pressure in the geological formation in order to prevent dangerous blowouts. An equally important role of drilling fluids is to form a mud cake around the walls of the borehole to prevent collapse prior to the casing installation.[6, 9] The introduction of drilling fluids, while a considerable advantage for production, has resulted in the most significant waste stream from a drilling operation, thus requiring careful management and judicious reuse and disposal methods to be developed.

Drilling fluids have evolved from simple solutions of clay and water to the sophisticated synthetic mixtures used today. Drilling fluids are commonly grouped into two divisions, water-based and oil-based fluids. Although the manufacturing of drilling fluids has become a science, with many companies developing unique recipes that are frequently customized to meet the needs of an individual formation, there are several ingredients that are commonly found in most fluids.[1, 9] Barite ($BaSO_4$), a heavy, highly dense mineral, is added to most fluids as bulk to ease the pressure caused by drilling. Clay, most commonly bentonite, is a significant component of the drilling fluid because it is chemically neutral and typically will not react with the formation (see Table 2.1).

Table 2.1 Components Are Added to Drilling Fluids to Provide the Following[6]

Physical Property	Purpose & Effect
Viscosity	Ability to circulate freely through drilling equipment and borehole
Stability	Maintenance of physical and chemical aspects under high temperature and pressure common to boreholes Ability to remain in suspension (no settling out)
Durability	Ability of chemical composition and physical characteristics to withstand cleaning processes of fluid circulation system (no degradation)
Environmental compatibility	Minimization of impact on geologic formation and ease of disposal
Cost-effectiveness	Cheap availability because they are used in very large quantities

Drill Rig Configuration

The majority of modern drilling operations involve the use of rotary drilling technology. However, drill rig configurations differ depending upon whether drilling will occur onshore or offshore. Onshore drill rigs are typically large land-based platform structures, but may consist of smaller truck-mounted rigs in rough terrain. Offshore drilling involves a floating drill platform that may or may not be connected to a ship.

2.2   Development

The developmental stage involves preparations for full-scale oil production. This begins once the exploratory drilling program has confirmed the presence of commercial quantities of hydrocarbons. As with exploration, development occurs in stages of intense research.[6, 8] The first stage of development is a feasibility study to determine the economic practicality of the project, and includes an assessment of resource size and waste disposal procedures (cost and types). A development team, typically consisting of geologists, engineers, economists, and numerous other professionals, is assigned the task of assessing every aspect of the future oil-producing region. Often, additional data are needed to completely analyze the project, so the exploratory drilling program may be expanded. This may consist of additional geophysical work and supplementary field exploration programs. The result of the feasibility study is a field development plan for oil production.[1, 6] Figure 2.5 outlines the developmental stages of conducting a feasibility study.

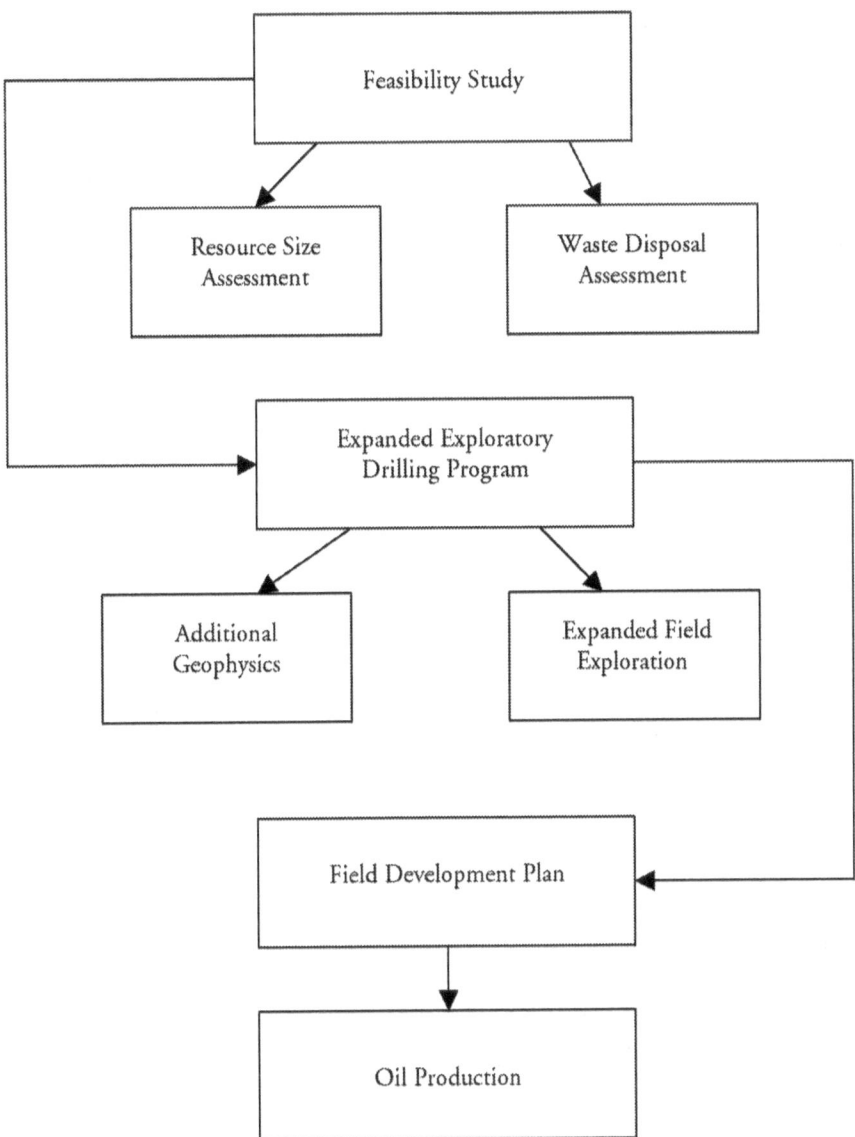

Figure 2.5   Schematic of Developmental Stages

2.3 Oil Production

Oil production is the ultimate result of a successful exploratory drilling program. Once oil is discovered and found to meet the minimum criteria for commerciality, a decision has to be made whether to hook up the discovery well and begin production, or whether the newly discovered field has to be appraised to properly define the field limits and volumes of hydrocarbons. The appraisal process involves drilling one or more wells and acquiring as much engineering data as possible from those wells. Following a successful appraisal, it may be decided to drill more "infill" wells, which have shorter spacing between them than the appraisal wells. These infill wells are called development wells and are necessary to increase reservoir drainage and to achieve more efficient recovery of hydrocarbons. The process of drilling appraisal and development wells may be accomplished in several phases over many years. Completion of each phase is an opportunity to evaluate results, integrate newly acquired data, and update the field development plan to better ensure the success of subsequent phases. It may also be an opportunity to secure the budget for the next phase.

Under the field development plan, intense efforts are made to get the field ready for production, which may include the construction of pipelines and flow-stations, and the prior environmental assessment work associated with them.

When the decision is made to start production, any remaining drilling equipment is removed, a production wellhead is installed, and the wells are connected to a flowstation. In drilling terminology, this phase is referred to as well completion.

2.3.1 Well Completion

Well completion begins with the removal of the drilling equipment from the borehole. This must be preceded by cementation of the casing that had been placed into the borehole during the exploratory drilling stage. The upper end of the casing is joined at the surface to the wellhead, or "Christmas tree," a system of pipes and valves used to control the flow of oil during the production phase.

In order for oil to flow through the casing and into the well, a perforating gun loaded with explosive charges is placed at the bottom of the well, and the charges are detonated.[8] In addition to perforating the well casing, the charges also penetrate into the formation to create a pathway for oil recovery. Once perforation has occurred, production tubing is placed into the well. A sealing device called a production packer is placed at the bottom of the production tubing to regulate pressure within the well during oil production.

2.3.2   Pumping Operations

In some formations, natural pressure is sufficient to push the oil to the surface. However, pumping, or natural lift, is more common.[1, 6] Pumps are designed to pass through the production tubing and rest at the bottom of the well. Pumps either function through a rotary spinning or a vertical plunging movement. A large pumping unit located at the surface provides the power to the pumping operation.

2.3.3   Primary Stimulation/Production Enhancement

Prior to installing the pumping apparatus in the well, it is often necessary to take additional steps that will stimulate primary oil recovery. Stimulation involves the use of chemical or physical techniques to create or enhance openings for oil to pass through the formation and into the well for extraction to the surface.[8] These techniques are described below.

Fracturing

Oil migrates through naturally occurring fractures. For oil production to be efficient, it is often necessary to increase the size and number of fractures to ease the flow. In some formations, the initial fractures to the formation caused during well perforation are sufficient to enhance oil flow. However, it is more likely that additional physical force must be used.

Fracturing works by pumping pressurized fluid, either water or oil, down the well to forcefully produce cracks in the formation. The fluid is mixed with proppant materials—hard, durable material such as sand or ceramic pellets that remain in the formation cracks after the fluid dissipates to continue to prop open the fractures and create an open pathway for oil production.[8]

In an effort to reduce waste and avoid contamination issues, it is becoming more common to use water instead of oil in the fracturing process.

Acidizing

In readily dissolvable formations such as limestone or dolomite, fractures may be enhanced or created chemically using the acidizing technique of fluid stimulation.[6, 8] Pressurized acid such as hydrochloric acid (HCl) is injected into the formation through the production tubing. The acid enters the formation through perforations and dissolves the surrounding rock to create larger openings for the oil to pass through. Hydrochloric acid (HCl) is most commonly used, as it easily dissolves limestone and dolomite and is readily available.

Coiled Tubing

Coiled tubing is a recent technology that was developed for horizontal drilling from already-existing vertical wells.[8] The flexibility of the method has allowed for adaptation to well completion and production enhancement.

Large reels of seamless high-pressure steel tubing are trucked onto the site and threaded into the well through the wellhead. The tubing is equipped with a drill bit to penetrate through rock. The coiled tubing advances through the rock in a manner similar to that of rotary drilling, using a rotating bit and drilling fluid. In addition to creating large, uniform conduits for oil flow, coiled tubing has advantages in terms of waste management in that it only uses drilling fluids, thus it is more economical than fluid-based stimulation. Table 2.2 summarizes the various techniques used for primary stimulation and production enhancement.

Table 2.2   Primary Stimulation Techniques[8]

Technique	Explanation/How It Works
Newly Installed Wells	
Fracturing	Pressurized fluid is pumped down the well to forcefully produce cracks (fractures) in the formation The fluid is combined with proppants—hard material such as sand or ceramic pellets—that continue to prop open the fractures after the fluid disperses/dissipates
Acidizing	Acids are pressure-injected into the formation through the production tubing Acids enter the formation through perforations and dissolve the surrounding rock to create larger openings for oil to pass through. Hydrochloric acid (HCl) is most commonly used, as it easily dissolves limestone and dolomite—common formation rocks.

Technique	Explanation/How It Works
	Already-Producing Wells
Coiled tubing	Large reels of seamless high-pressure steel tubing are threaded into the well through the wellhead
	They are equipped with drill bits to penetrate through rock formations to create fractures
	The system functions similarly to rotary drilling except with the ability to drill horizontally
	It is a favorable option for previously drilled areas that may still contain oil that was previously inaccessible
	There is less waste to manage than with other techniques—coiled tubing uses drilling to create fracture instead of fluids

2.3.4 Oil Recovery/Recovery Enhancement

Oil recovery typically occurs in three stages that are based on an increasing use of technology. Primary oil recovery begins after stimulation efforts are implemented,[6] (as described in Section 2.3.3). Secondary and tertiary recovery use well injection to introduce pressurized materials into the formation to further enhance oil production.[6, 8] In order to accomplish this, injector wells are drilled within a radius of the production well. Well injection is carefully monitored, as the introduction of materials into the subsurface is a waste management challenge.

<u>Secondary Recovery Methods</u>

Secondary recovery involves the injection of high-pressure flows of water or natural gas into the formation. The injected material supplements the natural formation pressure to force oil from the rock. Water injection, or water flooding, is most commonly used, as it is highly efficient and readily available for use in most locations. It is necessary in situations in which the natural aquifer is weak and cannot provide sufficient pressure to drive oil through the rock. Use of natural gas injection is limited to reservoirs with structural problems, such as steeply dipping or highly permeable layers, so that gravity can assist with the process.[11] Natural gas has a lower viscosity and a higher degree of mobility than oil and commonly dissipates without enhancing recovery unless structural control (a geologic subsurface structure that controls the natural flow direction of oil and gas, such as a dome) is present in the formation. The use of natural gas for well injection is limited because of high cost and limited application.

Tertiary/Enhanced Recovery Methods

Tertiary recovery methods are implemented after primary and secondary methods are no longer economically feasible. Methods such as miscible flooding and chemical application are most suited to the recovery of light oil.[6, 8] Miscible flooding uses the injection of materials that are capable of mixing completely with oil and withstanding separation—such as liquid natural gas or carbon dioxide—to decrease viscosity and increase the ability of the oil to pass through the formation rock. Chemicals such as polymers or surfactants may be injected to create a highly viscous emulsion that is easily recovered from the formation.

Tertiary methods also may be used when secondary techniques have failed. Thermal processes, which apply heat to the formation, are most successful in recovering heavy oil.[8] Heat reduces the viscosity of heavy oils, allowing for easier flow through the formation. The stages of oil recovery, the associated techniques used at each stage and process descriptions are summarized in Table 2.3.

Table 2.3   Summary of Oil Recovery Stages

Recovery Stage	Technique	Process Description
Primary	Stimulation	■ Fracturing ■ Acidizing ■ Coiled tubing
Secondary	Well Injection	■ Water flooding ■ Natural gas (with gravity)
Tertiary (Enhanced)	Enhanced Well Injection	■ Miscible flooding ■ Chemical injection ■ Thermal application ■ Microbial application

2.4 Refining of Crude Oil

The crude oil extracted from the ground during the production stage is a mixture of different types of hydrocarbons and non-hydrocarbon residual elements such as sulfur, oxygen, and nitrogen. In order to use the energy from the hydrocarbons, they must be processed, or refined, into petroleum products such as gasoline, diesel fuel, and fuel oil.

Basic refining begins in the oil field. Oil produced at the wellhead typically contains gas and water, which must be removed prior to shipping to the refinery. This is accomplished by pumping the oil from the production wells to a gathering station, a field facility that contains separation equipment.[12] First, oil and gas are separated in large cylindrical tanks in which the oil settles to the bottom and the gas rises to the top. The oil and gas are then pumped into separate systems for water removal. This water is called production water and is a major source of waste that must be appropriately contained and managed.

In the refinery facility, refining begins with physical separation to extract the different petroleum types, or fractions, from the crude oil.[12, 13] The refining process continues with chemical conversion to manipulate the fractions into different products, and ends with the purifying stage, in which the different fractions are further refined into specialized products.

2.4.1 Physical Separation

Physical separation uses distillation to divide the hydrocarbons into different types, or fractions, based on the differences in boiling point range.[12] The process begins with atmospheric distillation, in which crude oil is entered into a fractional distillation column to extract the lighter hydrocarbons. The heavier residual material is sent to a vacuum distillation column to extract the heavier hydrocarbons.

2.4.1.1 Fractional Distillation

Fractional distillation is the most important part of the refining process as it is able to separate crude oil into many different components using very subtle differences in boiling points.[12, 13] The process begins when crude oil is heated in a boiling unit, or furnace, which is usually powered by high-pressure steam generated from the gas byproducts of oil production. As the crude oil is boiled, most substances reach the vapor phase and enter the fractional distillation column, a long cylinder that is fitted with perforated trays.

The vapor cools as it rises through the column. When the cooling vapor reaches the layer of the column at which the temperature matches its boiling point, it condenses to form a liquid. The liquid fractions either pass from the distillation column to condensers for further cooling and then to storage tanks, or to other units for further refining through chemical conversion processes.[1, 12] The heavier residual materials settle to the bottom of the column and are transferred to the vacuum distillation unit for additional refining. Figure 2.6 is a schematic diagram of this process.

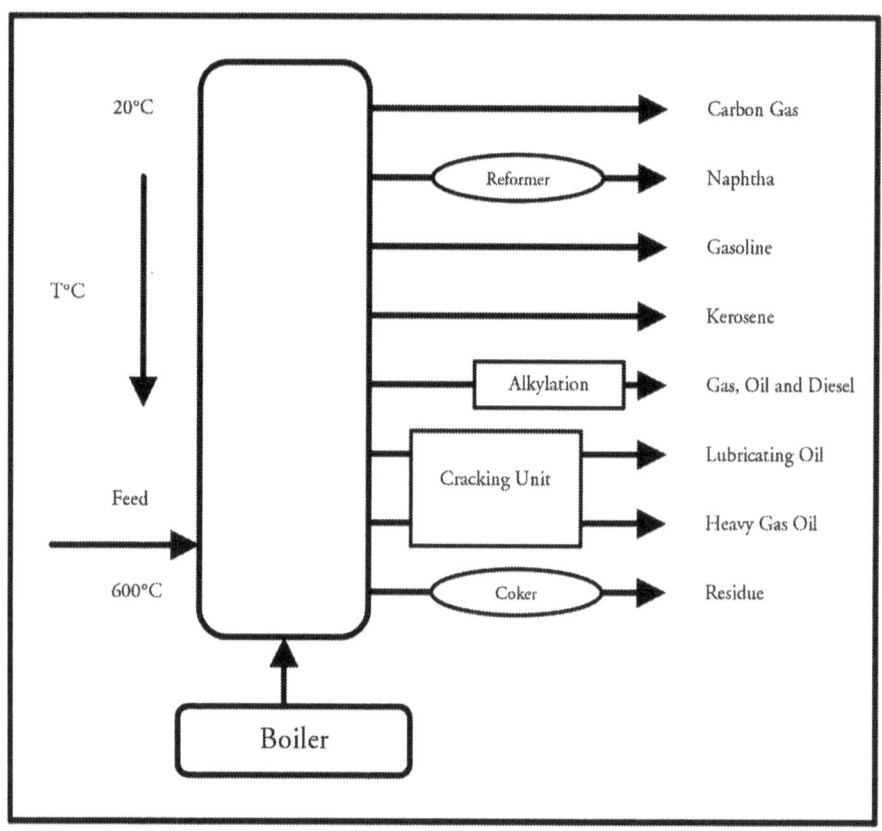

Figure 2.6   The Fractional Distillation Column[10]

2.4.1.2   Vacuum Distillation

Vacuum distillation is a process that separates heavy residue into usable fractions.[12] This process would normally require very high temperatures to reach the boiling points of the heavier hydrocarbons in the residue. But by using low pressure in the vacuum column, the boiling points are lowered, and the process is able to use significantly less energy. Vacuum distillation increases the efficiency of the refining process through lower energy requirements and results in maximum hydrocarbon recovery from crude oil, which is important as a waste minimization and reduction effort.

2.4.2 Chemical Conversion

After fractional and vacuum distillation is completed, the hydrocarbon fractions are subject to chemical conversion to produce different products by changing the molecular size or structure of the hydrocarbons.[12, 13, 14] Chemical conversion is accomplished by one of three basic methods: decomposition (division) separates larger hydrocarbon molecules into smaller ones; unification mixes smaller hydrocarbon molecules together to produce larger ones; and alteration (rearrangement) produces different fractions by manipulating the molecular structure.

2.4.2.1 Decomposition

Large hydrocarbons from the distillation process are decomposed, or divided, into smaller hydrocarbons by thermal or catalytic cracking methods.[14] Thermal cracking uses heat to separate large hydrocarbons at high temperatures. Catalytic cracking incorporates a catalyst, a substance that accelerates the decomposition process. The hydrocarbon fractions that result from the cracking process are circulated through a fractional distillation column to produce lighter hydrocarbon fractions.

The most commonly used thermal cracking methods are visbreaking and delayed coking. Visbreaking reduces the viscosity of heavy hydrocarbons to produce lighter, more viscous products such as tar. Delayed coking uses very high temperatures to remove hydrogen from the heavy residues to form a heavy, carbon-rich residue called coke, which is used as a fuel source for industrial plants.[12, 14] The coking process is described as "delayed" because it occurs in large drums where the heavier materials settle out, instead of occurring directly in the furnace where it is heated.

Catalytic cracking introduces a catalyst to accelerate the cracking process. Fluid catalytic cracking uses a fluid catalyst and a high temperature to crack heavy hydrocarbons into gasoline and diesel products.[14] Hydrocracking uses hydrogen gas as a catalyst, under lower temperature and higher pressure, to produce liquid fuel products such as gasoline and kerosene.

2.4.2.2   Unification

Unification, or combining, refines smaller hydrocarbons into larger ones. The most commonly used methods are alkylation and polymerization.[13, 14] Alkylation combines an acid and a catalyst with lighter hydrocarbon fractions to produce high-octane hydrocarbons for gasoline blending. Polymerization unites smaller molecules from the same hydrocarbon fraction to form larger molecules that are used to produce gasoline fuels.

2.4.2.3   Alteration

Alteration rearranges the molecular structure of one hydrocarbon fraction to produce another fraction, most commonly through isomerization or catalytic reforming. In isomerization, molecular structures are rearranged to produce compounds of the same family, called isomers. The resulting isomers may possess markedly different physical and chemical properties, which can result in the production of valuable compounds from less-valuable source compounds.[12] Catalytic reforming uses a catalyst to rearrange low-octane gasoline into more desirable higher-octane gasoline, and is mainly employed to increase the anti-knock capacity of the gasoline. The three chemical conversions discussed above are summarized in Table 2.4.

Table 2.4   Summary of Chemical Conversion Refining Methods

Method	Types	Process Description
Decomposition (Dividing)	Thermal cracking	Visbreaking—uses heat to reduce viscosity and to break heavy fractions into more usable, viscous products Delayed coking—converts residue into usable fractions of petroleum (with lower boiling points) and into coke
	Catalytic cracking	Fluid catalytic cracking—fluids help to crack heavy fractions into lighter fuels Hydrocracking—uses hydrogen and lower pressure for cracking
Unification (Combining)	Alkylation	Lighter hydrocarbons are mixed with an acid and a catalyst to produce high-octane hydrocarbons for gasoline
	Polymerization	Smaller molecules of the same hydrocarbon fraction are combined to form larger molecules; commonly used to produce fuels

Method	Types	Process Description
Alteration (Rearranging)	Catalytic reforming	Uses a catalyst to rearrange low-octane fuels into more desirable higher-octane fuels
	Isomerization	Molecular structures are rearranged to produce compounds of the same family (isomers)

(Adapted from Speight, James G. *Handbook of Petroleum Analysis*).[14]

2.4.3 Finishing/Purifying Stage

Prior to leaving the refinery, hydrocarbon fractions are further modified to produce final products with specific properties. This is accomplished by additional chemical or physical treatment and by blending, in which fractions are mixed with additives and other components.[1, 13] Some common physical and chemical finishing methods are summarized in Tables 2.5 and 2.6.

Table 2.5   Summary of Physical Finishing Methods

Method	Process Description
Absorption	Separation based on either molecular size or molecular type Uses differences in solubility
Solvent extraction	Separation based on molecular type Uses differences in miscibility (mixing capacity) with introduced component
Crystallization	Separation based on either molecular size or molecular type Uses differences in melting point and solubility
Adsorption	Separation based on either molecular size or molecular type Uses capability of hydrocarbon to adhere to porous materials

(Adapted from *The Petroleum Handbook*, Shell Corporation, 1983)

Table 2.6   Summary of Chemical Finishing Methods

Chemical Method	Process Description
Hydrogen treating (Hydrotreating)	Use of hydrogen as a catalyst to remove impurities Most often used to remove hydrogen sulfide (H_2S) and called hydrodesulfurization
Sweetening	Fractions produced from crude oil with high sulfur content have a very strong odor Odor is removed by treatment with caustic aqueous solutions
Sulfur recovery	Used to remove sulfur from gas fractions Sulfur is an economic byproduct of refining and is sold to make fertilizers and pharmaceuticals
Clay treatment	Fine clays applied to mineral oil fractions to improve color and stability

(Adapted from *The Petroleum Handbook*, Shell Corporation, 1983)

2.5   Distribution

Distribution involves the transportation and storage of refined petroleum products to enable use by the consumer. Transportation is divided into two main categories: either by large-capacity vehicles such as tanker trucks and ships; or by pipeline.[1] The mode of transportation selected depends on the type and quantity of product and the distance between the refinery and the consuming market. Storage methods used are dependent on the quantity and characteristics of the refined products.

2.5.1   Transportation

Since the world's supply of oil tends to be located in specific pockets of the globe that are isolated from the majority of the oil-consuming market, transportation is a crucial part of the oil industry. In terms of capacity, ocean-bound oil tankers comprise the majority of the oil transportation sector, followed by rail cars, and lastly by tanker trucks.[15] The most publicized episodes of oil spillage have resulted from incidents involving transportation, which have in turn increased public awareness of the need for transportation methods that prevent oil and associated waste products from entering the environment.

2.5.2 Pipelines

Pipelines are used for transportation when the output from a refinery is sufficient to support a concentrated market area. For example, in Canada, most of the main oil fields are located in remote areas of the western part of the country. To reach the main consumer markets in the southeastern part of Canada and in the northeastern United States, a large pipeline network was built.[8] The initial cost of pipeline construction was eclipsed by the cost savings associated with the elimination of conventional and archaic methods of transportation.

2.5.3 Storage

The type of storage selected is dependent on the capacity of oil products to be stored, and on their subsequent usage.[1, 9] Large quantities of petroleum products may be stored in aboveground storage tanks (ASTs)—which are commonly grouped in large open areas and referred to as tank farms—or in underground storage tanks (USTs). Smaller quantities may be stored in smaller ASTs or USTs, drums, or other containers.

REFERENCES

1. Royal Dutch Shell Group Staff. 1983. The Petroleum Handbook, 6th Edition. Amsterdam: Elsevier Science. 710.

2. Daniel Yergin. 1999. The Prize.

3. Samuel T Pees. 2003. Oil History. www.oilhistory.com

4. Robert Gaston. 2001. Discovery of the Spindletop Oilfield, www.drillinginfo.com/wonderings20000101.html

5. Karl Dyk. 1961. Geophysical Surveys. Chapter 11. In: Moody, Graham B. Petroleum Exploration Handbook. New York: McGraw-Hill.

6. F. Jahn, M. Cook and M. Graham. 1998. Hydrocarbon Exploration and Production. Developments in Petroleum Science, Vol. 46. New York: Elsevier Science. 384.

7. Von Buelow. EU Essential Points in Use of Geophysics. Oil & Gas Journal, Special Issue: 100 Years of Industry Leadership. Vol. 100, Iss. 35, Aug. 2002. Pages 56–66.

8. Robert Bott. 1999. Exploring Canada's Oil-and-gas industry, 6th Edition. Calgary, Alberta: Petroleum Communication Foundation. 101.

9. F. K. North. 1985. Petroleum Geology. Winchester, Massachusetts: Allen & Unwin Inc. 607.

10. Craig C Freudenrich. 2003. How Oil Drilling Works. www.howstuffworks.com/oil-drilling.htm

11. Natural Gas Supply Association, The. 2003. www.naturalgas.org

12. H. S. Bell. 1959. American Petroleum Refining, 4th Edition. Princeton, New Jersey: Van Nostrand Company. 538.

13. Craig C Freudenrich.2003. How Oil Refining Works. www.howstuffworks.com/oil-refining.htm

14. James G Speight. 2002. Handbook of Petroleum Analysis. Hoboken, New Jersey: John Wiley & Sons Inc. 409.

15. Jean-Paul Rodrigue. 2004. Transport Geography on the Web. Hofstra University Dept. of Economics and Geography. http://people.hofstra.edu/geotrans

3.0 WASTE CHARACTERISTICS

Each stage of operation associated with the oil industry generates specific types of waste as shown in Table 3.1 below. In order to plan effective and adequate waste management strategies, it is necessary to identify and understand the waste characteristics of each operational stage.

The oil industry has recognized the need to be a good environmental citizen, and has been at the forefront in developing innovative methods to minimize the impact of the industry's activities on the environment by reducing the amount of waste generated.[1]

Guidelines similar to those established by the American Petroleum Institute (API) have been adapted by the industry throughout the world to effectively deal with waste. Many of these guidelines are applicable to the oil industry waste streams summarized in Table 3.1.

Table 3.1   Summary of Waste Characteristics

Process or Stage	Waste Characteristics
Exploration	
Accessing target locations Geophysics field programs	Road construction—removal of vegetation and topsoil Line cutting—vegetation removal for instruments Waste from vehicles and/or aircraft used to transport crew and equipment, and to perform geophysical surveys (fuel, tires, oil, filters, etc.) Seismic surveys—spent explosive charges (if used)
Base camp facilities	Removal of vegetation and topsoil Building materials from camp construction Household debris, food waste Wastewater and sewage Fueling and equipment-maintenance waste
Drilling and Production	
Drill rig operation & maintenance Well production & completion	Drilling fluids (may be oil-, water-, or synthetic-based) Rock cuttings (shale, limestone, sandstone, etc.) Produced water—consists of naturally occurring water from geologic formations, brines, and injection water from enhanced recovery techniques Air emissions due to flaring of gas Other wastes Used oil, oil filters, hydraulic fluids, lubricants, and solvents Solid wastes—drums, containers, trash Wastewater—equipment cleaning and runoff Used spill kit material—absorbent pads, booms Rigwash Paints, paint wastes, sandblast waste Scrap metals

Process or Stage	Waste Characteristics
Oil Refining	
Refinery facilities	■ Air emissions Gases from combustion of fuel and waste incineration (SO_2, NO_x) ■ Particulates ■ Gas flaring ■ Liquid waste ■ Water used for refinery processes (process water)—may contain hydrocarbons, chemicals, slop-tank liquids Standard wastewater (sewage) Corrosion and hydration inhibitor Solid waste Sludges remaining from liquid waste treatment and tank bottoms Used catalyst materials Paints, paint waste, sandblast materials Paraffins, scale Mercury from manometers and pressure and flow gauges Noise generated during refinery operations
Distribution and Storage	
Transportation	Large-scale impacts (oil tankers at sea, train cars and tanker trucks on land) Smaller-scale impacts (pipeline breaks/leaks) Compressor and meter station wastes
Storage	Leaks and spills from tanks Water and sediment accumulation in storage tanks ("tank bottoms")

3.1   Waste Associated with Exploration

Oil exploration involves three major stages: accessing target locations, establishing a base camp for programs of long duration; and executing a geophysical exploration program. Each stage has specific waste concerns that are addressed with careful planning in order to manage waste through minimization.[2]

3.1.1   Accessing Target Locations

Most targets for oil exploration are located in isolated rural areas, posing a challenge in terms of accessibility. In these situations, it is often necessary to construct access roads to bring in equipment and personnel. Road construction involves the removal of trees, vegetation, and topsoil—actions that have the potential to cause problems of soil erosion and natural habitat disruption. Solutions to these potential problems focus on minimizing the disruption of the area, in order to produce less waste (see Table 3.2). To reduce soil erosion, cleared vegetation is used as part of an erosion-control plan, and disturbed topsoil is stockpiled for replacement after the exploration program ends.[2] Replanting the area with native plant species further minimizes the impact on the natural habitat.

Table 3.2   Summary of Remote Access Issues

Potential Problem	Solution
Soil erosion	Use of cleared vegetation for soil conservation and erosion-control measures
	Stockpiling of disturbed topsoil for future replacement
Habitat impact	Replanting of investigation area with plant species native to area

(Adapted from API Guidance Document E5: Waste Management in E&P Operations)

3.1.2   Geophysics Field Programs

Geophysical exploration programs are typically executed on a large scale, either from land, air, or sea. In order to adequately gather data from the target area, the program must be efficiently planned and executed. As technology has advanced to improve the efficiency of geophysics, additional benefits have been recognized in terms of waste generation, as streamlined processes result in less waste by focusing on minimization.[1, 2] Some of the key environmental and waste management issues involved with geophysical exploration are summarized in Table 3.3 below.

Table 3.3   Summary of Geophysical Exploration Issues

Location	Geophysical Survey	Waste Issue	Solution
Land	Gravity	Disturbance of vegetation and topsoil	Avoidance of removing root structures of plants to promote re-growth Leaving topsoil in place
Sea	Seismic	Detonation of explosives Soil erosion, wildlife disturbances	Newer technology that uses non-explosive sources—pressurized-air guns in water and vibroseis trucks on land
Air	Magnetic	Operation and maintenance of aircraft	Managing aircraft from central base, typically at nearest airport, with proper aviation fueling and repair facilities

(Adapted from API Guidance Document E5: Waste Management in E&P Operations)

Land-based gravity and seismic surveys require vegetation removal in order to place parallel lines of measuring instruments on the ground surface. This specific type of vegetation removal is called line cutting. Leaving plant root structures intact to promote regrowth and exercising care to leave topsoil in place minimizes waste from line cutting.[2] Seismic surveys also require the detonation of a charge to propel energy waves through the earth. Newer technology uses non-explosive sources—such as pressurized-air guns in water and vibrating vibroseis trucks on land—to generate the energy waves, instead of the powerful explosives that were more commonly used in the past.[3] Non-explosive sources provide the powerful energy needed without disrupting aquatic or terrestrial life or triggering soil erosion problems.

Magnetic surveys are performed from the air, using satellites or aircraft. When air surveys are used, the majority of waste is associated with the aircraft and is dealt with at a centralized location equipped for aviation fueling and maintenance issues.[2, 3] In sensitive and remote environments, airborne survey techniques are preferred to improve accessibility and preserve natural resources.

3.1.3   Base Camp Facilities

In remote areas where intensive geologic mapping and geophysical exploration programs are planned, a base camp may be established near the target location.

Base camps typically consist of accommodations for the workforce, such as dining and showering facilities, fueling stations for vehicles and aircraft, and an equipment maintenance area. Since the base camp must be self-sufficient in terms of waste disposal and water treatment, generated wastes are minimized and are dealt with mainly by recycling and reuse[2] as shown in Table 3.4.

Table 3.4   Summary of Base Camp Wastes and Issues

Component	Waste Produced	Solution
Accommodations for workforce	Building materials	Use of modular units that can be broken down and transported off-site Recycling of materials
Dining and showering facilities	Food wastes	Composting
	Household trash/debris	Incineration Containerization for shipment to disposal facility
	Wastewater and sewage	Implementation of collection systems—either septic tanks or filtration systems using permeable earth-covered beds to treat waste without affecting drinking-water supply
Fueling and equipment maintenance areas	Waste oil, oil filters, fuel	Collection and transportation off-site to recycling facility Performing fueling operations in an appropriate manner
	Fuel containers	Collection and transportation off-site to recycling facility

(Adapted from API Guidance Document E5: Waste Management in E&P Operations)

3.2   Waste Associated with Drilling and Production

The largest source of waste generated in the oil industry is the drilling and production phase as shown in Figure 3.1. With the advent of rotary drilling and the subsequent introduction of drilling fluids, significant increases in production and drilling, as well as in waste generation.[2, 4] Additional sources of waste include produced water, which occurs naturally with crude oil in most formations, and must be extracted prior to the refinery stage. Atmospheric emissions from the operation of diesel-powered drilling equipment and the flaring up of gases also must be considered.[3] Other sources of waste include equipment and pipeline maintenance and subsurface geological testing.

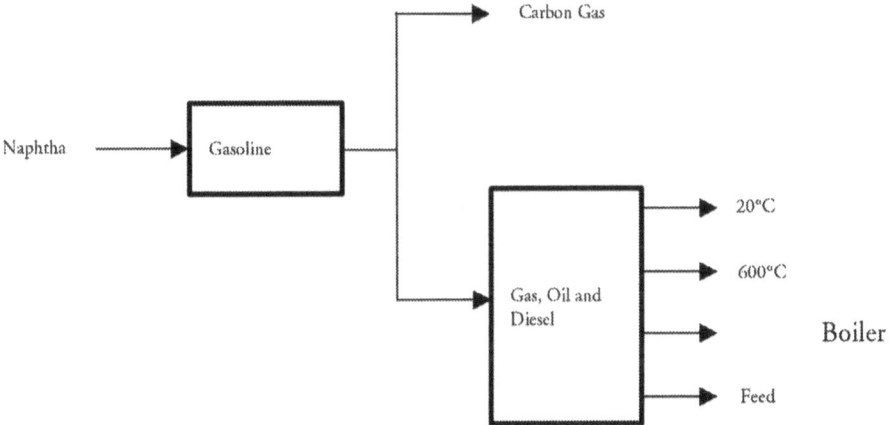

Figure 3.1   Schematic of Drilling Operations and Associated Waste Streams

3.2.1   Drilling Fluids and Cuttings

Drilling fluids are an integral part of the drilling process. They are used to lubricate the drilling equipment, to cool the drill bit, to regulate pressure inside the borehole to prevent dangerous blowouts, and to flush rock cuttings from the borehole.[5] Drilling fluids originally consisted of simple water-and-clay mixtures. As drilling technology has advanced, fluids have become increasingly complex mixtures that are specifically engineered for the conditions in a given drilling formation.[6]

Drilling fluids are grouped into two main divisions: water-based and oil-based fluids. In addition to oil or water, three essential components are found in drilling fluids: solids such as bentonite clays for viscosity and thickness; minerals such as barite ($BaSO_4$) for weight; and numerous chemical additives to enhance drilling performance.[7]

Oil-based fluids were introduced for offshore drilling, in order to help penetrate through great depths of rock under significant heat and pressure.[7, 8] Replacing water with oil products such as diesel fuel or mineral oil produced a stronger, more cohesive fluid that would not disintegrate, in contrast to water-based fluids.[8]

However, oil-based fluids posed significant problems in terms of disposal. Unlike water-based fluids, the chemical composition of oil-based fluids does not allow for disposal through on-site discharge.[7] Water-based fluids, composed mostly of water with clay and minor, inert chemical additives, could be left on-site to integrate back into the environment. To prevent environmental damage, oil-based fluids and associated rock cuttings must be disposed of at special facilities, thereby increasing the waste generated.[7, 8]

The creation of synthetic-based drilling fluids was a major achievement in terms of waste management.[1, 8] Synthetics are produced through chemical manipulation of pure organic materials such as fatty acids and olefins, and are characterized by low toxicity, allowing for disposal through on-site discharge, thereby eliminating the need for special off-site disposal measures.[10] Additionally, synthetic-based fluids are durable enough to replace oil-based fluids in special drilling conditions and can be reused, cutting down on the quantity of waste generated during drilling. Some of the benefits of using synthetic-based fluids are discussed in Table 3.5.

Table 3.5   Environmental Benefits of Synthetic-Based Fluids

Versus Oil-Based Fluids and Water-Based Fluids	
Versus Oil-Based Fluids	Versus Water-Based Fluids
Lower concentration of potential contaminants—due to lack of hydrocarbons	Less waste produced—synthetics can be adjusted for reuse at other drilling operations
Cuttings can be safely discharged into the environment	Faster drilling time—reduces energy consumption and air emissions
Increased worker health and safety from lower toxicity	Increased drilling control—synthetic fluids are engineered for specific borehole conditions
Reduced air pollution—fluids are discharged on-site and do not require transport to special disposal facilities	Smaller drilling footprints—synthetics enable directional drilling from a single well, instead of requiring multiple wells
Reduced landfill usage and disposal costs from cuttings discharge on-site	

(Adapted from U.S. Department of Energy, Environmental Benefits of Technology)

In economic terms, although synthetic-based fluids cost more initially, reuse and the elimination of special disposal requirements result in savings.[1, 7, 8]

Produced Water

Produced water is extracted, or produced, from the subsurface during oil production. This category includes formation water naturally occurring within the geologic units; brines, which are highly saline and mineralized; and injection water from the tertiary/enhanced oil recovery stage.[3, 4] Most formation water is saline and usually contains trace elements and metals, dissolved hydrocarbon fractions, and suspended solids. Injection water may contain chemical additives such as coagulants, corrosion inhibitors, detergents, emulsifiers, and surfactants, all necessary for improving the recovery process.

Produced water may also contain naturally occurring radioactive materials (NORMs) that leach out from clay-based hydrocarbon formations such as shale and sandstone.[9] The main radionuclides of NORMs are radium-226 and radium-228, produced through the decay of uranium and thorium isotopes. Dissolved radium is most often found in highly saline formation waters. Under appropriate conditions, dissolved radium may precipitate with trace elements such as barium, strontium, or calcium to form sulfate, carbonate, or silicate scales or sludges, posing significant problems for equipment.[6, 9]

Especially during later phases of oil extraction, produced water is second only to drilling fluids and cuttings in terms of waste volume.[1, 4] As an oil well reaches the end of its production life span, the amount of oil extracted decreases while the amount of produced water increases, becoming an increasingly larger issue for waste management.[7, 8]

Other Wastes

Maintenance and support of the drill rig and associated equipment is another significant source of waste materials. Waste includes used oil and oil products such as filters and lubricants, solid waste such as containers and construction materials, waste water from rig cleaning (rigwash) and storm water runoff, and used spill kit materials[1, 6] as shown in Table 3.6. For these types of waste, the most appropriate waste management technique is to minimize usage as much as possible, and to recycle materials that must be used.

An additional significant source of drilling and production waste is from atmospheric emissions generated by diesel-powered equipment and gas flaring.[3] Techniques have been pioneered in the oil field to make use of gases that would normally be flared by turning them into a feasible fuel source for power generation. Cleaner-burning fuels to power drilling equipment are becoming more viable.

Table 3.6   Other Drilling and Development Wastes

Waste Type	Examples	Solutions
Used oil products	Used oil Filters Lubricants and greases Solvents Cleaning rags	Containerization for off-site recycling Installation of reusable filters Minimization of waste generation by keeping equipment in optimal condition
Water	Rig cleaning (rigwash) Stormwater runoff	Containment systems and/or catch basins incorporated into rig layout
Solid wastes	Construction materials Containers Pipeline maintenance	Recycling or disposal at appropriate facility Buying materials in bulk to minimize trash produced
Spill-related waste	Used spill kit materials—absorbent pads, booms	Containerization for off-site recycling Use of good housekeeping skills to prevent spillage
Air emissions	Drill rig operations Gas flaring	Switching from diesel to cleaner-burning fuels Reuse of gases that would be flared as energy sources to minimize waste

(Adapted from API Guidance Document E5: Waste Management in E&P Operations)

Waste Associated with Oil Refining

In general, refinery waste consists of air emissions, liquids, solids, and noise.[10] Specific waste characteristics may vary by the size of the refinery, the extent of maintenance operations, and the quantity and nature of crude oil processed at the facility.

3.2.2   Air Emissions

Air emissions consist of gases—mainly sulfur dioxide (SO_2) and nitrogen oxide (NO_x)—and particulates. The main sources of air emissions from oil refineries are combustion from the burning of energy sources for fuel, incineration of wastes, and the flaring of excess gases.[11] A minor source is the venting of hydrocarbon vapors from processing equipment and storage tanks.

Oil refining is the use of heat to transform crude oil into usable hydrocarbon products, which requires a significant amount of energy. Many refineries produce power by burning diesel fuel in large generators. In an effort to become more efficient in terms of crude oil usage and waste reduction, refinery residuals such as gases or heavy oils may be burned to generate power.[12] Sulfur dioxide and nitrogen oxides are released into the air from combustion from energy production and gas flaring. Particulates are introduced into the air through the incineration of solid wastes such as residual hydrocarbons, and through catalytic refining processes.

3.2.3   Liquid Waste

Liquid waste consists of processed water that may be impacted by petroleum constituents and other chemicals; condensates, the liquid residuals from cooled hydrocarbon vapors; and wastewater such as sewage and other non-impacted water.[2]

Oil refining is a water-intensive process, as water is required as a coolant for heat-intensive processing applications. To minimize this waste stream, refinery operations are designed to reduce water consumption by continuously reusing water, and by selecting refining processes that do not require large amounts of water.[11, 12] To further reduce water usage, air may be used in place of water as a coolant.

Refinery-process water and condensates may contain volatile compounds, hydrocarbons, dissolved organics, and suspended solids. The beginning stages of treatment use oil/water separators to remove hydrocarbons, filtration systems to remove solids, and biological treatment with microorganisms and aeration to remove organics. The remaining sludges are included in the solid waste category.[2, 5] Standard wastewater such as sewage and non-process water is kept separate from refinery-process water and is disposed of in water treatment plants.[5]

3.2.4   Solid Waste

Solid waste consists of fine suspended particles removed from crude oil, impurities from liquid waste treatment, and spent catalysts from catalytic refining procedures.[2, 5] Additional solid wastes include refuse and debris from standard business operations.

Solid particles removed from crude oil and the solids remaining after water treatment processes are commonly referred to as "sludges." Treatment of sludges consists of dewatering to remove the liquid component, and disposal.[12] In the past, it was common for refineries to incinerate dewatered sludge waste. However, increasing consciousness of air emissions and escalating costs have resulted in the use of landfill disposal methods.[5]

Catalytic refining procedures introduce a substance to accelerate the refining of heavier crude oil components. The most commonly used catalysts contain metals such as platinum and palladium, and metal oxides such as aluminum oxides, chosen for durability to allow for maximum reuse in the refining process.[5, 12] At the end of their life spans, these metallic catalysts are processed to remove valuable metal components and are disposed of with sludge residues through landfilling or incineration. A summary of the categories of oil refinery wastes and their characteristics is described in Table 3.7.

Table 3.7   Characteristics of Oil Refinery Waste

Basic Categories of Oil Refinery Waste		
Category	Waste Source	Characteristics
Air	Fuel burning	Refineries are typically powered by combustion of a fuel source
	Emissions	Vapors—venting from valves and tanks within the refinery Particulates—from burning fuels
	Flaring	Combustion of excess gases—typically hydrogen sulfide (H_2S)
Liquid	Condensates	Liquid waste from cooled vapors
	Process water	Used in hydrocarbon processing—may contain volatiles, hydrocarbons, dissolved organics, and solids in suspension
	Waste water	Sewage and non-process water waste
Solid	Crude oil residues	Fine suspended particles
	Water treatment	Solid and/or sludge remainders Naturally occurring sediment in water supply
	Spent catalysts	Generated during catalytic cracking and hydrocracking procedures
Noise	Equipment	Noise from refinery operations—may affect worker safety and quality of life of surrounding neighbors

(Source: The Petroleum Handbook, Shell Corporation, 1983)

3.2.5 Noise

Noise output is a serious concern because of the potential impact on worker safety and the quality of life of neighborhoods and wildlife near the refining facility.[3, 5] Noise is classified as a waste because it is possible to implement design features while constructing a refinery in order to reduce or eliminate sound levels. According to the Shell Corporation Petroleum Handbook, Shell refineries are constructed with the goal of eliminating noise output by surrounding the central operations with sound-blocking features such as vegetation and large storage tanks.[5]

3.3 Waste Associated with Distribution and Storage

Distribution involves the transportation of petroleum products from the refinery to the retail market. It is closely associated with storage, as products are disbursed from bulk facilities. Waste generated at the distribution stage is highly variable and difficult to characterize in a general manner.[2] The beginning of the transportation phase reduces the degree of waste control that is inherently part of oil production and refining, and introduces a host of random variables that require comprehensive planning. Waste characteristics of distribution and storage are best discussed on a large scale and a small scale to cover the spectrum of possible occurrences.

3.3.1 Transportation

Large-scale transportation involves oil tanker ships at sea, and tanker trucks and rail cars on land, and has the potential to create major waste from spills. To prevent waste, these large-scale transportation vehicles are designed to prevent spillage by incorporating details such as lined tanks and pressure-control valves.[2, 5, 6] Additional waste generated from vehicle operation and maintenance, such as spilled fuel and used oil and filters, is minimized by recycling and careful use of resources.

Smaller-scale transportation includes pipelines, compressor and meter stations for pumping the petroleum product in the pipeline, and bulk fuel trucks. Waste includes leakage and spillage caused by pipe breakage or pipe maintenance, and is also managed by design. Pipelines are constructed with liners and are equipped with electronic sensors that are carefully monitored by a computer to detect and prevent spills.[3] Pipeline maintenance is performed by a technique called pigging, which involves sending a device called a pig through the lines to dislodge blockages. Recent technological advances have produced "smart" pigs, which are equipped with sensory equipment to detect potential leakage zones as part of preventive maintenance measures.[3] If a spill or leak does occur, additional waste will be produced from cleaning, which may include absorbent materials or contaminated soil that must be excavated and treated. Compressor stations contain large petroleum-fueled engines connected to piping systems and other associated components such as slop tanks, cooling towers, and generators. Compressor-station wastes are mainly derived from maintenance operations.

3.3.2   Storage

Storage of refined oil products can occur at the refinery or after transportation to bulk fuel facilities. Oil products are stored in large tanks, located either above ground (ASTs) or underground (USTs). Storage facilities range in size from large facilities with numerous tanks—called tank farms—to the smaller-capacity underground tanks found at commercial fueling stations.

Leakage and spillage from tanks are the typical wastes that result from the storage of oil products. An additional form of storage waste is found on tank bottoms, and consists of sediment that has settled out from the oil products during storage.[5] Leakage and spillage occur during the transfer of oil from tanks, and can range in magnitude from small surface spills to widespread underground contamination from faulty retail gasoline stations. Waste-prevention measures such as double-walled tanks and leak-detection devices have been implemented at most storage-tank facilities. Storage tanks are regularly inspected and cleaned to remove tank-bottom materials, which are treated and disposed of in methods similar to those used for solid waste sludge generated from oil refining processes.[5, 11, 12]

REFERENCES

1. United States Department of Energy, Office of Fossil Energy. 1999. Environmental Benefits of Advanced Oil and Gas Exploration and Production Technology. Washington.

2. American Petroleum Institute. 1997. Environmental Guidance Document: Waste Management In Exploration and Production Operations, 2nd Edition. Washington DC: American Petroleum Institute. 78.

3. Robert Bott. 1999. Exploring Canada's Oil-and-gas industry, 6th Edition. Calgary, Alberta: Petroleum Communication Foundation. 101.

4. F. Jahn, M. Cook and M. Graham. 1998. Hydrocarbon Exploration and Production. Developments in Petroleum Science, Vol. 46. New York: Elsevier Science. 384.

5. Royal Dutch Shell Group Staff. The Petroleum Handbook, 6th Edition. 1983. Amsterdam: Elsevier Science. 710.

6. ICF Consulting. 2000. Overview of Exploration and Production Waste Volumes and Waste Management Practices in the United States. American Petroleum Institute. 112.

7. Stanislav Patin. (translation by Elena Cascio). 1999. Environmental Impact of the Offshore Oil-and-gas industry. Chapter: Waste Discharges. www.offshore-environment.com/discharges.html.

8. Jonathan Wills. 2000. Muddied Waters: A Survey of Offshore Oilfield Drilling Wastes and Disposal Techniques to Reduce the Ecological Impact of Sea Dumping. Sakhalin Environment Watch. http://www.offshore-environment.com/jonathanwills.html

9. K. P. Smith. 1992. An Overview of Naturally Occurring Radioactive Materials (NORM) in the Petroleum Industry. United States Department of Energy, Office of Domestic and International Energy Policy. Argonne, Illinois: Argonne National Laboratory.

10. Marshall Sittig. 1978. Petroleum Refining Industry: Energy Saving and Environmental Control. Energy Technology Review No. 24, Pollution Technology Review No. 39. Park Ridge, New Jersey: Noyes Data Corporation. 374.

11. Stanley E. Manahan. 1997. Environmental Science and Technology. Boca Raton, Florida: Lewis Publishers. 641.

12. James G. Speight. 2002. Handbook of Petroleum Analysis. Hoboken, New Jersey: John Wiley & Sons Inc. 409.

4.0 ENVIRONMENTAL ISSUES OF THE OIL INDUSTRY

The primary environmental issues associated with the oil industry involve the potential risks to human health and the environment through air, groundwater, surface water, and land contamination. Typically, contamination enters the environment through emissions, releases, or spills, which may be generated during each stage associated with the oil industry—from exploration and production to refining and transportation of oil products to the consumers. Other potential areas of concern include noise and thermal pollution as well as the impact to sensitive areas from physical disturbances involved with activities such as equipment mobilization and construction.

4.1 Air-Quality Issues

Air contamination with respect to the oil industry occurs directly through emissions and gas flaring, and indirectly through altering the chemical balance of the atmosphere, which results in acid rain and ozone pollution.[1] Clean air is crucial to sustain the health of humans, animals, and the environment, and air quality is addressed through preventive measures involving technological solutions to reduce or eliminate harmful contaminants.

4.1.1 Atmospheric Emissions

The main sources of atmospheric emissions associated with the oil industry are related to the combustion of energy sources for fuel, incineration of wastes, and the flaring of excess gases from collection and refinery systems.[2] Atmospheric emissions consist of gases, particulates, and hazardous air pollutants (HAPs).[3] The main gases emitted from the above processes are sulfur oxides (SO$_x$), nitrogen oxides (NO$_x$), and carbon monoxide (CO). Particulates are small particles that directly enter the air as a result of vehicle engine exhaust and industrial combustion processes, or as a result of chemical reactions. Also of concern are HAPs from the internal combustion engines of vehicles and industrial facilities such as refineries[2, 3] that are toxic to human health and the environment.

Sulfur Oxides

Sulfur occurs naturally in crude oil and in coal. When these fuel sources are burned for energy, or when gasoline extraction occurs at an oil refinery, sulfur oxide gases (SO$_x$) are formed.[2, 3] Sulfur dioxide (SO$_2$) is the most significant of the sulfur oxides, as it is highly mobile in the environment. It readily dissolves in water to form highly corrosive sulfuric acid (H$_2$SO$_4$), which affects the structural integrity of buildings and monuments—especially those constructed of easily dissolvable calcium carbonates (CaCO$_3$)—as well as the foliage of plants, and the acidity of freshwater lakes.[3] It has a very strong, noxious odor and is a respiratory irritant for humans and animals. Additionally, sulfur dioxide is easily adsorbed to particulate matter in the atmosphere, and is a significant cause of acid rain.[3, 4]

Technology to reduce or eliminate atmospheric emissions of sulfur dioxide has been developed for use in the oil industry and in secondary sources such as coal-fired power plants.[5] Sulfur recovery methods have been instituted at many oil refineries to remove the sulfur content from fuels. Coal-fired power plants are equipped with technologies such as injection beds of fine-grained calcium carbonate in the boilers, and injection of lime slurries into the gas emissions column in order to eliminate SO$_2$ at the source of combustion prior to discharge.[5]

Nitrogen Oxides

Nitrogen oxides (NO$_x$) are generated by the combustion of hydrocarbon fuels at high temperatures in the presence of elevated oxygen concentrations.[6] Nitric oxide (NO) is the primary nitrogen oxide produced, and is quickly converted to highly mobile nitrogen dioxide (NO$_2$) in the atmosphere. While NO is colorless and odorless, NO$_2$ is reddish-brown and has a distinct odor at elevated concentrations. Similar to sulfur dioxide, nitrogen dioxide is easily incorporated into water vapor and particulate matter, and contributes to acid rain.[5, 6]

Nitrogen oxides combine with volatile organic compounds (VOCs) in the presence of heat and sunlight to form ground-level ozone (O$_3$), a major contributor to urban smog.[1, 6] VOCs are produced from the evaporation of hydrocarbon fuels and hydrocarbon-based products such as solvents and chemicals. Trace quantities of VOCs also occur naturally in the atmosphere as byproducts of photosynthesis.

Ground-level ozone and nitrogen oxides are serious threats to human health, causing respiratory-system damage and breathing difficulties, especially to young children and the elderly.[1] The environmental impacts of acid rain include severe damage to plant tissues and damage to soil and water.

Emissions of nitrogen oxides have been greatly reduced by the advent of emission-control systems in modern motor vehicles.[6] Devices incorporated into the exhaust system are designed to lower the temperature of combustion, thereby depriving the system of conditions necessary for nitrogen formation.

Carbon Monoxide

Carbon monoxide (CO) is formed when carbon-based fuels are burned at reduced levels of oxygen, and are key emissions from the internal combustion engines of motor vehicles.[1, 2] Although small amounts of CO occur naturally in the atmosphere, the majority is introduced from traffic congestion in urban areas. CO is colorless, odorless, and is a significant risk to human health, as it is easily absorbed into the bloodstream and rapidly decreases the ability of red blood cells to convey oxygen throughout the body.[1]

Technology to reduce carbon monoxide emissions is incorporated into modern vehicles through the addition of catalytic converters.[2] Prior to entering the atmosphere, vehicle exhaust passes through the catalytic converter, which contains a catalyst—usually a metal such as platinum or palladium—which rapidly converts carbon monoxide into less-harmful carbon dioxide and water through the process of oxidation.

Particulates

Particulate matter consists of small particles suspended in the air, and includes dust, dirt, ash, smoke, soot, and liquid vapor. The main sources of particulate emissions from the oil industry are from the combustion of hydrocarbon fuels as a power source for refineries, in the internal combustion engines of vehicles, and from the incineration of wastes.[1, 2, 3] Particulates are either emitted directly into the air or formed through chemical reactions between gaseous emissions and water vapor in the atmosphere.

Particulates affect human health through the respiratory system. Suspended in the air, particulates are breathed into the lungs, where they are absorbed into the bloodstream and lymphatic system. They are then transported through the body to damage organs and tissues. Of special concern are cancer-causing compounds called polycyclic aromatic hydrocarbons (PAHs), emitted by internal combustion engines and easily absorbed into the human body.[1] Particulates impact the environment by decreasing visibility, causing aesthetic damage to buildings and monuments through the deposition of soot and ash, influencing weather patterns, and polluting water, air, and soil.[1, 2] As well, the small particles provide adhesion surfaces for atmospheric water vapor, forming smog and promoting acid rain generation.

Technological controls to reduce or eliminate particulate emissions into the atmosphere are engineered into most processes associated with the oil industry.[2, 5] During fuel production, sedimentation processes are used to remove particles with the assistance of gravity, producing cleaner fuels.[2] Facilities used for waste incineration or power generation through the burning of coal are equipped with air filters to remove particles from the emissions prior to discharging them into the atmosphere.

Hazardous Air Pollutants

Hazardous air pollutants (HAPs) are substances that are toxic to both human health and the environment.[3] HAPs may cause cancer, birth defects, reproductive system impairment, and other serious health effects such as damage to the respiratory, immune, and nervous systems. Sources of HAPs are typically man-made and include vehicular emissions of benzene and toluene from gasoline, and the burning of fuels and waste containing metals such as mercury and lead.[1, 13] Recognition of the toxicity of HAPs to human life and to the environment has prompted the implementation of strict regulatory guidelines in many countries, and the development of technology solutions within the oil industry.[4]

Lead is the HAP that poses the greatest concern as an atmospheric emission, as it is highly toxic to living organisms.[1] Lead accumulates in body tissues and causes severe damage to the nervous system. Lead is highly resistant to corrosion, and was once a common additive to gasoline and paints. However, reformulation of gasoline and the elimination of lead in paint has significantly decreased lead levels in the atmosphere.

Mercury, a highly mobile and toxic heavy metal, is produced as vapor during the incineration of coal and certain types of waste.[1, 12] Elemental mercury is commonly found in manometers, switches, and flow meters at meter and compressor stations. Mercury is highly toxic when inhaled. It causes further damage when it is easily absorbed into soil and water, where it is ingested by organisms. Through bioaccumulation, it has the ability to cause cumulative damage throughout the food chain.[2] Efforts to prevent mercury contamination are incorporated into maintenance, refining, and air treatment processes.

4.1.2   Gas Flaring

A typical oil deposit contains a mixture of natural gas and crude oil. Natural gas containing significant amounts of hydrogen sulfide (H_2S) is referred to as "sour gas" and is a major issue in terms of both human safety and environmental impact. Hydrogen sulfide is highly toxic. It is very dense and can be fatal to humans, animals, and plant life, causing oxygen deprivation if released into the atmosphere.[1] The process of flaring uses heat to convert deadly H_2S into lighter and less-toxic sulfur dioxide and carbon dioxide emissions.[2, 5]

Gas flaring is an air-quality issue, as it does not completely eliminate air emissions. It merely converts a highly toxic substance into SO_2 and CO_2, which are less toxic but equally damaging as air pollutants.[3] The strong odor and loud noise associated with flaring operations have impacted the environment and led to complaints from communities near facilities that flare. Recent studies in the oilfields of Canada have illustrated that flaring may not result in complete combustion, and is thought to emit volatile organic compounds (VOCs) into the atmosphere.[7]

In an effort to improve air quality, the oil industry has implemented several approaches. If gas is present in sufficient quantities, flaring may be an issue of wastefulness and inefficiency that is preventable by implementing facilities to process the sulfur from sour gas as an economical byproduct of oil production.[2] When flaring cannot be avoided, new techniques such as re-injection into the formation or using the gas to power small electricity-producing generators may be implemented.[1]

4.1.3   Acid Rain

Acid rain is produced when waste emissions such as sulfur dioxide and nitrogen dioxide react with water, oxygen, and sunlight in the atmosphere to produce acidic compounds. These acidic compounds, mainly sulfuric acid and nitric acid, enter the hydrologic cycle and are deposited into the water and soil.[1] Winds can carry these acidic compounds over great distances, which may turn a regional pollution problem into a larger problem.

Acid rain can impact the environment by acidifying of lakes and streams, and by killing. It also corrodes structures and buildings, especially those constructed out of calcium carbonate, which is highly sensitive to acid. The effects of acid rain on lakes and streams are most noticeable in sensitive areas where soils do not contain sufficient compounds to neutralize, or buffer, the impact of acid raises the pH of the water and causing damage to aquatic habitats.[5] Soils that do not contain sufficient buffering compounds typically release aluminum into the environment, thereby increasing the toxic effects of acid rain and thus affecting more plants. However, in highly alkaline waters, acid rain may reduce pH levels, thereby improving aquatic habitats.

Excessive nitrogen content in acid rain can cause oxygen depletion in lakes and streams.[2, 5] Decreased oxygen content in water bodies promotes the growth of algal blooms, which can be toxic to aquatic life and can inhibit plant growth. Excess nitrogen affects humans and wildlife by affecting drinking water and food sources from fish and shellfish. Acid rain may pose additional risks to human health in individuals with respiratory illnesses and breathing difficulties due to the presence of sulfur and nitrogen compounds.

Efforts implemented by the oil industry to control acid rain have focused on reducing emissions of SO_2 and NO_2.[3, 6] Emission-control devices on vehicles and refining facilities, as well as improvements in the efficiency of refining processes, have significantly reduced emissions at the source.[2] Conversion to cleaner fuel sources such as natural gas and lighter hydrocarbon derivatives have also decreased the threat of acid rain.

4.2   Groundwater Issues

The main source of fresh, or potable, water on Earth is found in underground formations of porous rock called aquifers. Water is produced during the hydrologic cycle—as precipitation falls from the sky and evaporation occurs from surface water and vegetation. Water that enters an aquifer by filtering through the ground—a natural mechanism to remove impurities—is called groundwater. Groundwater aquifers fulfill demands for clean drinking water and irrigation for agriculture. Groundwater contamination caused by the oil industry is a result of introducing contaminants such as oil products, metals, brine, and dissolved constituents into the environment through spills, leaks, or emissions during production or waste disposal operations.[1]

4.2.1   Contamination Sources

Although oil occurs naturally in the subsurface with water, contamination problems arise during production. Secondary oil recovery methods such as fracturing produce wider pathways for oil to migrate into areas of the reservoir that may be used as clean water sources.[2] Enhanced recovery methods, which mainly focus on the injection of a stimulation compound such as steam or water, often introduce chemicals into the reservoir that have the potential to migrate.

Produced water—the combination of naturally occurring formation water and injection water introduced during secondary and enhanced recovery—and drilling fluids also have the potential to affect groundwater, particularly during disposal. Most produced water is highly saline and contains significant concentrations of dissolved metals. Depending on the chemical composition of the formation rock, produced water may also contain naturally occurring radioactive materials (NORMs) such as radium, which is commonly associated with saline formation water.[1,2] In the past, produced waters were discharged onto the surface to be reabsorbed into the ground. However, recognition of the potential contamination issues posed by produced waters have led to the recycling of produced waters for various drilling procedures, and the usage of on-site treatment facilities.[2]

Other potential contaminants can be introduced during drilling or refining. Potential contaminants from drilling include solvents, oil products, and greases that are used as part of drill-rig operation and maintenance, and can be introduced to the subsurface in stormwater and rigwash-water runoff. To prevent this, the drilling platform is surrounded by catch basins from which water is channeled to on-site treatment facilities.[2] Oil refineries prevent discharging contaminants by constructing water treatment and recycling facilities, and by employing diligent spill-control measures and technologies.

4.2.2 Receptors

Once contamination enters the groundwater, the primary receptors are drinking-water and irrigation-water aquifers, and the people and livestock and crops that depend on them as a source of potable water.[1,2] The effects of groundwater contamination that are not immediately stopped and remediated are most noticeable in these primary receptors (i.e., humans, plants, animals) when the contamination affects their health or, in extreme situations, causes their death.

4.3   Surface-Water Issues

Surface water is comprised of oceans, lakes, rivers, and streams, and is a significant part of the hydrologic cycle. It directly influences water quality for aquatic habitats, human resources such as food supplies from fish and shellfish, and water-based industries such as fisheries and tourism. Surface-water issues are primarily focused on the impact of oil contamination to watersheds and drainage areas through migration patterns.

4.3.1   Oil Migration

Oil spilled on water quickly spreads across its surface to form a slick. This oil slick is highly mobile and subject to rapid dispersal and migration through the effects of hydrodynamic flow, dispersion, convection, and wind and water currents on surface water bodies.[1, 2]

Short-term responses to surface-water oil spills consist of immediate containment and recovery with absorbent booms and surface-skimming devices. Long-term solutions to oil spills include the natural weathering processes of evaporation, emulsification, dissolution, and biodegradation. Oil occurs naturally in the environment as a result of the decay of microorganisms. Natural weathering eventually reduces the oil into compounds such as CO_2, H_2O, and microbial biomass, which are all reincorporated into the environment.[1, 2, 5]

4.3.2   Receptors

The primary receptors of surface-water contamination from oil spills are the aquatic life found in watersheds and drainage areas, and sensitive areas such as swamps, estuaries, and floodplains.[1] Although the immediate impact of an oil spill is serious and subject to intense media sensationalism surrounding short-term effects such as oil-covered birds and animals, these sensitive areas are naturally equipped with bacteria and organisms that decompose oil, thereby reducing long-term effects of spills.

4.4   Land/Soil Issues

Soils are critical to sustaining life on Earth. They provide the necessary natural resources for agriculture, industry, and terrestrial organisms. Contamination of soils through spills or releases, solid and hazardous waste disposal, and leachate from landfills are serious primary causes of environmental issues. However, once contaminants are introduced into the soil, they can also cause serious secondary problems through evaporation into the atmosphere and drainage into groundwater and surface water.

4.4.1   Contamination Sources

Spills and releases of contaminants with the potential to impact soils are found in each stage of the oil industry, from exploration and production to refining and transportation. At the exploration and production stage, spills and releases occur during drilling accidents or pipeline transmission of crude oil, and are initiated by equipment failure, human error, or natural disaster. The implementation of technology such as blowout-prevention valves on drill rigs and computerized sensors on pipelines have been used successfully as spill-prevention methods. Additionally, technology has eliminated a large number of spills related to human error by mechanizing many tasks previously performed by people. Spills and releases from refineries are mainly attributed to leaking storage tanks and associated piping, and are prevented by engineering mechanisms such as double-walled tanks and catch basins for pipes. In instances when preventive measures fail and spills and releases occur, affected soils are excavated for remediation treatment or disposal.

If hazardous wastes generated by the oil industry cannot be recycled or reused, they are disposed of at facilities that are specially equipped to handle the waste. Disposal is the least favorable option for waste handling, as it is the most costly and inefficient, and requires special procedures such as containerization in tanks and drums. Hazardous waste disposal is a highly technical and specialized industry whose mandate is to safely manage waste while minimizing environmental impact and risks to human health.

Solid wastes that cannot be recycled, reused, or incinerated on-site are disposed of through burial at a landfill. Contamination from landfills occurs when water infiltrates and dissolves materials in the waste, producing a noxious liquid called landfill leachate. Leachate can migrate through soil and into groundwater. To prevent contamination of surrounding soils and subsequent migration of leachate into groundwater, landfills are lined with protective, impermeable materials and are equipped with systems to collect leachate from the landfill for separate treatment.

4.4.2 Receptors

The primary receptors of soil contamination are people, wildlife, livestock, and crops, vegetation, and sensitive ecological habitats.[1, 2] People and animals could be directly exposed to contaminated soil by either ingestion or body contact, and indirectly by eating plants and crops that absorb contaminants as they grow. Plants and crops are also directly affected by spills and releases of contaminants that can cause death or severe structural damage to their systems. Sensitive ecological habitats such as mangrove swamps in Nigeria, estuaries on the coasts of America, and Arctic tundras contain rare and delicate flora and fauna that can be severely damaged or destroyed by soil contamination. Once contaminants enter into the soil, they have the potential to infiltrate groundwater, surface water, and the air, affecting the environment and human health on a very large scale with potentially severe consequences.

REFERENCES

1. Laura Jones, Fredericksen Liv and Wates Tracy. 2002. Critical Issues Bulletin: Environmental Indicators, 5th Edition. Vancouver: Fraser Institute. 136.

2. James A. Fay and Dan S. Golomb. 2002. Energy and the Environment. New York: Oxford University Press. 314.

3. United States Environmental Protection Agency. Air Quality: Six Common Air Pollutants. http://www.epa.gov/air/urbanair/6poll.html

4. United States Environmental Protection Agency. Clean Air Markets: Environmental Issues. http://www.epa.gov/airmarkets/envissues/index.html

5. Stanley E. Manahan. 1997. Environmental Science and Technology. Boca Raton, Florida: Lewis Publishers. 641.

6. United States Environmental Protection Agency, Office of Air Quality Planning and Standards. 1998. NO_x: How Nitrogen Oxides Affect the Way We Live and Breathe. Research Triangle Park, North Carolina.

7. Robert Bott. 1999. Exploring Canada's Oil-and-gas industry, 6th Edition. Calgary, Alberta: Petroleum Communication Foundation. 101.

5.0 ENVIRONMENTAL REGULATIONS

Environmental regulations that govern waste management for the oil industry are best discussed in terms of onshore industries and offshore industries. Onshore industries involve individual countries, which regulate waste nationally or federally, and sometimes at state or provincial levels. Offshore industries and the waste they generate are an international concern and are regulated by conventions and agreements that are established regionally between countries that share jurisdiction over water bodies, or on a global scale, by organizations such as the International Maritime Organization (IMO) and the United Nations (UN).

In general, most oil-industry environmental regulations have been developed and enacted because of the resulting impact of oilfield activities. The earliest laws were designed to protect oil resources from the overzealous extraction and development that plagued the first major oil finds and resulted in massive oil depletion and various impacts to soil, vegetation, and nearby water systems. Modern laws have become more proactive, and are designed to avoid the potential for environmental impact and risk to human health. From the past to the present, these laws have included, either directly or indirectly, the basic components of waste management, focusing on the smart usage of resources and using methods that minimize waste generation and environmental impact.

5.1 Onshore Industry—National Laws

The oil industry is active on nearly every continent of the globe. Regulations to protect the environment, its natural resources, and its inhabitants are established by national or federal government authorities. Additional regulations—on a state, provincial, or territorial level—may also be implemented to manage the unique resources and local issues that may be associated with the activities of the oil industry in these areas.

Regulations are in constant flux, subject to frequent revisions based on improvements in technology and an increasing desire of governments and citizens to maximize the protection of the environment and associated natural resources. In every aspect of the oil industry, from exploration and development to production and refining, a continuous review of legislation is crucial to success, both of operations and public relations. For the sake of brevity, this section will examine environmental regulations that are currently enforced in two significant oil-producing countries, the United States (U.S.) and the United Kingdom (UK).

5.1.1 Comparison of National Laws: United States vs. United Kingdom

A country-by-country discussion of specific laws pertaining to the oil industry and associated waste management stipulations is beyond the scope of this document. Instead, to illustrate the similarities between the environmental legislation in different nations, it is useful to compare the policies of the two largest oil-producing Western countries, the United States and the United Kingdom. In the United States, regulations are under the jurisdiction of the Environmental Protection Agency (EPA). Establishment of regulations in the United Kingdom is much less streamlined, as they are imposed by the European Commission (EC) and are enforced locally by different agencies such as the Environmental Agency (EA) in England and Wales, and in Scotland by the Scottish Environmental Protection Agency (SEPA). Nationally, regulations are enforced by the Department of Food, the Environment, and Foreign Affairs (DEFRA) or the Department of Trade and Industry (DTI).

Regardless of the differences between enforcing entities or the consequences imposed for violations, laws are structured similarly in both nations, and address key phases of the oil industry. In order to provide a brief overview of these regulations, the discussion will focus on exploration and production, transportation, spills, hazardous and non-hazardous waste, and guidelines pertaining to groundwater and surface-water impacts, air-quality issues, and spill prevention and control. Examination of selected regulations will provide a snapshot of applicable concerns pertaining to the oil industry.

Exploration and Production

Prior to starting an exploration program or bringing a newly discovered oil field into the production stage, comprehensive plans that focus on the appropriate legislation must be prepared. In terms of waste, both the UK and the United States have legislation requiring the implementation of a cohesive waste management plan in order to prevent the release of wastes into the environment, and to appropriately manage waste until disposal occurs.

Exploration and production activities in newly discovered areas typically require the preparation of an Environmental Impact Assessment/Environmental Impact Statement (EIA/EIS).[1] These reports analyze proposed activities to identify whether they have the potential to adversely affect human health or the state of the natural environment, and to provide proactive solutions for these problems. Additionally, in the UK, a license must be obtained prior to initiating any field activities.[2]

Transport

In both nations, the transport method of materials related to oil-industry activities is determined by whether the materials are hazardous or non-hazardous. Non-hazardous materials include oil, oil products, and non-hazardous waste. In the UK, the majority of transportation involving oil and oil products occurs through pipelines, and is governed by specific regulations based on the size and the use of the pipeline.[2] In the United States, shipments of materials are under the jurisdiction of the Department of Transportation (DOT) once they leave a private facility.[3] The DOT is also responsible for pipeline oversight.

Transportation of hazardous materials is strictly governed in both nations.[1,2] Prior to arranging transportation, the shipper must obtain permission for disposal of the materials at an approved facility. Special licenses must be obtained by shipping entities, and specific containers must be used to contain the waste. The regulations also contain a series of specific procedures that must be followed and documented throughout the shipping process, as well as during the disposal phase.

Groundwater

Regulations imposed by both nations focus on the recognition of groundwater as a valuable natural resource—as a source of drinking water for living creatures and a necessity for agriculture. Protection for underground aquifers is specified, specifically for those recognized as drinking-water sources. An equally important function of groundwater regulations in both the United Kingdom and the United States is to monitor the installation and operation of injection wells implemented during oil-drilling, as a waste disposal method.[1, 2]

Surface-Water Impact

Regulations pertaining to surface-water impacts are based on reducing waste at the source to prevent discharges, whether intentional or unintentional. Guidelines for waste reduction are a result of careful planning at each phase of development within the oil industry and by promoting energy conservation in an effort to reduce or eliminate impacts to surface water. In situations where surface-water discharges are allowed, the associated industry must demonstrate that all of the available procedures and technology are in place to prevent the release of potentially harmful materials into the environment.[2, 3] Individual regulatory agencies strictly enforce the use of permits for activities that are associated with waste discharges to surface water-bodies. These activities typically contain specific discharge limitations that must be achieved.

Air-Quality Issues

Environmental regulations in both nations pertaining to air-quality issues are based on the establishment of air-quality standards, or objectives to monitor emissions of concern to human health and the environment. Criteria have been established in both countries on a "not-to-exceed" basis for emissions commonly generated by standard operations and processes associated with the oil industry.[1, 2, 4] Regulated emissions include sulfur dioxide, nitrogen oxides, carbon monoxide, particulates, lead, and volatile organic compounds. Permits are often required and frequent inspections by regulatory agencies are commonly included as part of the regulatory process.

Spill Planning and Response

The serious impact of spills involving oil—whether during extraction, transportation, or refining—has led to the creation of strictly enforced regulations in both nations. In the UK, spill planning and response is included as part of the Pollution Prevention and Control Act of 1999, and focuses on the use of environmental management strategies, specifically Best Available Techniques (BATs) and Best Available Technology Not Entailing Excessive Cost (BATNEEC) as means to prevent spills.[2] Guidelines for BAT and BATNEEC are established by regulatory agencies and may be specifically adapted for individual operations within the oil industry. In the United States, the Oil Pollution Act was passed in 1990 as a direct result of the *Exxon Valdez* spill in Alaska. As a result of this incident, it was mandated that spill plans for a worst-case scenario must be developed prior to pursuing any of the stages of oil extraction.[5] Industry is also required to have adequate resources and appropriately trained personnel in place for agency approval of the spill plans.

Legislation requirements for the various stages of the oil-exploration process tend to vary slightly depending on where the exploration is taking place. Table 5.1 summarizes the various legal requirements in the United States as well as in the UK. The table outlines the differences and similarities in enforcement.

Table 5.1   Comparison of National Regulations Pertaining to the Oil Industry: United States vs. United Kingdom

Regulatory Issue	UK			USA	
	Legislation	Regulating Authority	Description of Legislation	Legislation	Description of Legislation
Exploration and production wastes	Petroleum Act, 1988	DTI	Requires a license for exploration, development, production and abandonment of all oil fields. All available methods and practices must be used to prevent waste discharge into the environment at all times.	Resource Conservation and Recovery Act (RCRA), 1976 Amended 1980	Guidelines for waste management. Classifies waste as hazardous (Subtitle C) or non-hazardous (Subtitle D). Requires submittal of Solid Waste Management Plans to EPA.
Environmental impact assessment/ Environmental impact statement	Town & Country Planning (England & Wales) Regulations 1999; Environmental Impact Assessment (Scotland) Regulations 1999; *Environmental protection act, Part I*	Local authorities	Established to assess the impact of future developments on the environment, prior to beginning development plans. Requires review and approval of local planning divisions.	National Environmental Policy Act (NEPA), 1969	Requires environmental impact assessments (EIAs) and environmental impact statements (EISs) for any actions that may adversely impact the quality of human health or the natural environment.

Regulatory Issue	UK			USA	
	Legislation	Regulating Authority	Description of Legislation	Legislation	Description of Legislation
Transport (Non-hazardous materials)	Pipelines Act, 1962; Works Regulations, 2000; Gas Act, 1986; Public Gas Transporter Pipeline Works Regulations, 1999	DTI	Oil transportation in the UK typically occurs by pipeline. Regulations are grouped according to the length and diameter of pipelines. All pipeline activities require environmental statements and are subject to regular reporting and inspections.	Department of Transportation (DOT) regulations for oil, oil products, and non-hazardous waste, enforced by EPA	Shipments occurring beyond the confines of a private facility are governed by the DOT.
Transport (Hazardous materials)	EC Regulation (259/93), Trans-frontier Shipment of Waste, 1994	DEFRA, EA, SEPA	Requires a special license for shipment and disposal of wastes.	Hazardous Materials Transportation Act (HMTA), 1975	Controls the shipment of hazardous materials, including hazardous wastes.
Pipelines	Pipelines Act, 1962, amended 2000	DTI	Regulates pipeline construction and monitoring to prevent and avoid environmental impact of activities.	Department of Transportation (DOT) regulations enforced by EPA	Pipelines are considered a mode of transport under United States law and are governed by specific DOT regulations.
Groundwater impact	EC Directive (80/68/EEC) and Groundwater Regulations, 1988	EA SEPA	To protect groundwater from discharges of contaminants. Includes regulation of injection wells. Permits are issued under the Water Resources Act of 1991.	Safe Drinking Water Act (SDWA), 1974	Establishes drinking water standards and regulation of injection wells. Laws to prevent contamination of underground sources of drinking water (aquifers).

Regulatory Issue	UK			USA	
	Legislation	Regulating Authority	Description of Legislation	Legislation	Description of Legislation
Surface water impact	Pollution Prevention & Control Act, 1999, and Regulations, 2000; *Environmental Protection Act, 1990 (Part I)	EA, SEPA, Local authorities	Mandates waste-minimization techniques to reduce the need for disposal and to conserve energy. Specifies returning sites to natural state. Discharges occur only after permits are granted.	Clean Water Act (CWA), 1972	Controls waste discharges into U.S. waters. Discharging operations must have National Pollutant Discharge Elimination System (NPDES) permits.
Air quality	Environment Act (Scotland), 1995; Air Quality Regulations, 2000; *Environmental Protection Act, 1990 (Part III)	Local authorities	Establishes air quality objectives for air pollutants of concern, including sulfur dioxide, nitrogen oxides, carbon monoxide, particulates, and lead. Requires air-quality improvement plans.	Clean Air Act (CAA), 1970, amended 1990	Establishes air-quality standards. Regulates emissions of sulfur dioxide, nitrogen oxides, carbon monoxide, particulates, hazardous air pollutants (HAP), and volatile organic compounds (VOCs).
Spill planning and response	Pollution Prevention and Control Act, 1999, and Regulations, 2000	EA, SEPA, Local authorities	Certain potentially polluting processes must be licensed by the authorities. Industries must employ environmental management using Best Available Techniques (BAT) and Best Available Technology Not Entailing Excessive Cost (BATNEEC).	Oil Pollution Act (OPA), 1990	On a worst-case-scenario basis. enforces spill planning and response Mandates national contingency planning and the creation of a cleanup fund. Stipulates limits of liability for spills. Assesses damage to the natural environment.

Regulatory Issue	UK			USA	
	Legislation	Regulating Authority	Description of Legislation	Legislation	Description of Legislation
Hazardous waste control/Disposal	Waste Management Licensing Regulations, 1994; *Environmental Protection Act, 1990 (Part II)*	EA SEPA	Hazardous wastes must be managed and disposed of using acceptable, environmentally sound procedures. Waste is classified as toxic, poisonous, explosive, corrosive, flammable, infectious, or ecologically toxic.	Resource Conservation and Recovery Act (RCRA), 1976, Amended 1980	Solid waste management. Classifies waste as hazardous (Subtitle C) or non-hazardous (Subtitle D). Requires submittal of Solid Waste Management Plans to EPA.
			Instituted regulations to protect the environment from industrial processes. Divided into nine parts, the Act governs impacts to land, water, and air. Relevant to the oil industry are:	Toxic Substances Control Act (TSCA), 1977	Allows EPA to evaluate and regulate chemicals that may have negative impacts on human health and the environment. Regulates identification and management of certain chemicals.
Hazardous waste cleanup			*Part I—Integrated Pollution Control by Local Authorities;* *Part II—Waste Management;* *Part III—Clean Air*	Comprehensive Environmental Response, Compensation and Recovery Act (CERCLA), 1980—commonly known as the Superfund program	Program established to identify sites that have released hazardous materials into the environment, enforces cleanup and evaluates damage to natural resources.

(Sources: UK Department of Trade and Industry website at www.og.dti.gov.uk; U.S.—API Guidance Document E5: Waste Management in E&P Operations)

As illustrated in the table, similar regulations have been established in the United States and the United Kingdom to manage the important environmental issues of the oil industry.

Hazardous Waste Control, Disposal, and Cleanup

Hazardous wastes are materials that pose serious, unacceptable risks to human health or the environment. In general, regulations in the UK and the United States have established a classification and identification system for all hazardous wastes. These regulations also require specific plans for waste management and disposal to be developed by the producer and the disposer. Procedures approved by regulatory agencies in both nations must be used during hazardous-waste control, management, and disposal. Regulations also exist in both nations to identify sites where hazardous materials have been released into the environment and to facilitate the cleanup and environmental restoration of these sites.[6] The main difference between regulations in the United States and the United Kingdom is that specific pieces of legislation have been enacted by U.S. regulators for key aspects involving hazardous wastes, while in the UK most aspects are included as part of the Environmental Protection Act.[1, 2]

5.2   Offshore Industry—International Agreements

In the offshore oil industry, waste management becomes an international concern due to the potential for impact on an environment that is shared by many nations, for economic and aesthetic purposes. Regulations are imposed on a global scale by organizations such as the International Maritime Organization (IMO) and the United Nations (UN), as well as on a regional scale between countries that share jurisdiction over water-bodies. Table 5.2 will outline the current global and regional agreements that are specifically relevant to the operations and procedures of the offshore oil industry.

Table 5.2   Summary of Recent Agreements— Global Offshore Oil Industry

Year Enacted	International Agreement Title	Description
1972	Convention on the Prevention of Marine Pollution by Dumping of Wastes and Other Matter (London Convention 1972) Under the authority of the International Maritime Organization (IMO)	To prevent dumping of wastes likely to pose hazards to human health or marine life, or to damage the natural resources and uses of the sea. Annexes to the Convention prohibit certain wastes—such as radioactive materials and industrial wastes—and require permits for dumping other wastes, such as sewage and formation waters.

Year Enacted	International Agreement Title	Description
1978	International Convention for the Prevention of Pollution from Ships, 1973, as modified by the Protocol of 1978 (MARPOL 73/78) Under the authority of the International Maritime Organization (IMO)	To eliminate pollution of the sea from ship discharges that may include oil, chemicals, or other harmful substances; to minimize accidental oil releases from ships and fixed or floating platforms; to improve pollution prevention and control from ships, with special focus on oil tankers. The 1978 Protocol declared that all ships must pass inspections to obtain an International Oil Pollution Prevention (IOPP) certificate prior to operating in waterways.
1982	United Nations Convention on the Law of the Sea (UNCLOS)	To establish and regulate the use and conservation of the marine environment and its natural resources. As recognized by the UN Conference on Environment and Development (UNCED) Agenda 21, UNCLOS serves as the internationally recognized legal standard for sustainable development and protection of the sea.
1992	United Nations Conference on Environment and Development ("Earth Summit of Rio")	A global environmental conference focusing on environmental protection and sustainable economic development. Resulted in Agenda 21—the Rio Declaration on Environment and Development, a set of guidelines promoting "eco-efficiency" through the following: ■ investigating alternative energy sources to replace fossil fuels; ■ emphasizing the use of public transportation systems to reduce vehicle emissions and associated air quality issues; ■ implementation of stricter controls on the production of toxic substances such as gasoline additives, and on the generation of hazardous waste

Year Enacted	International Agreement Title	Description
1996	Integrative Pollution Prevention Control (IPPC) Directive— European Union Directive 96/61/EC	Regulates the operations of large industrial facilities (such as oil refineries) by requiring companies to obtain permits from the appropriate regulatory bodies in EU countries. Granting of permits is determined by the use of 'Best Available Techniques' (BAT) to deal with emissions to air, water, and land, and promoting waste minimization and energy efficiency.

(Sources: The Fridtjof Nansen Institute's Yearbook of
International Cooperation on Environment and Development;
The United Nations online at www.un.org;
The European Union environmental policies online at
www.europa.eu.int/comm/environment.html)

5.2.1 Summary of Recent Agreements— Regional Offshore Oil Industry

Agreements, also called conventions, are also implemented regionally to enforce environmental regulations pertaining to the activities of the offshore oil industry.[7] A council of representatives from each member country is responsible for program management and regulation enforcement. In addition to following the stipulations of applicable regional agreements or conventions, each country must abide by international conventions if they have become signatories to the conditions of the convention. The UN also administers regional conventions in oil-producing areas across the globe.[8, 9, 10] For the sake of brevity, this section will focus on current regional agreements, which are summarized in Table 5.3.

Table 5.3   Summary of Selected Current Agreements—Regional Offshore Oil Industry

Year Enacted	Regional Agreement Title	Description
1992	Convention for the Protection of the Marine Environment of the North East Atlantic (OSPAR Convention), Paris, France Representatives from each member country create a governing body for program management and enforcement of regulations.	Includes all offshore oil-producing regions in Western Europe (Belgium, Denmark, Finland, France, Germany, Iceland, Ireland, the Netherlands, Norway, Portugal, Spain, Sweden, and the United Kingdom). The mandates of OSPAR are to protect the marine environment and human health, to restore affected environments whenever practical, and to prevent and eliminate pollution of the sea. Annexes to OSPAR specify the elimination of pollution from land-based sources, the prohibition of dumping radioactive waste at sea, and conservation of the biological diversity and ecosystems of the marine environment. OSPAR works in conjunction with the applicable UNEP regional programs and with the IMO
1992	Convention on the Protection of the Marine Environment of the Baltic Sea Area (1992 Helsinki Convention), Helsinki, Finland	To prevent pollution of the Baltic Sea area in order to promote environmental restoration and ecological balance. Annexes to the Convention stipulate the control of hazardous substances, the prevention of pollution from ships and onshore sources, and to prohibit waste-dumping
1974	United Nations Environmental Programme (UNEP) Regional Seas Programme Includes conventions for the following regions: 1. The Black Sea (Bucharest, 1992) 2. The wider Caribbean (Cartagena de Indias, 1983) 3. The East African seaboard (Nairobi, 1985) 4. The Kuwait region (Kuwait, 1978) 5. The Mediterranean Sea (Barcelona, 1976) 6. The Red Sea and the Gulf of Aden (Jeddah, 1982) 7. The South Pacific (Noumea, 1986) 8. The Southeast Pacific (Lima, 1981) 9. West & Central Africa (Abidjan, 1981)	To join coastal nations together in efforts to protect the world's coasts, inland waters, and oceans. Programs are adapted for the specific needs of each member nation but contain the following similarities: ■ An action plan for the management, protection, development, monitoring, and rehabilitation of coastal and marine resources ■ An intergovernmental agreement specifying general principles and obligations (may or may not be legally binding) ■ Detailed protocols for specific environmental problems such as oil spills The UN initially provides funding, with subsequent contributions from member nations. The program is administered by UNEP with cooperation from member nations

(Sources: The Fridtjof Nansen Institute's Yearbook of International Cooperation on Environment and Development online at http://www.greenyearbook.org; and the United Nations online at www.un.org)

5.3 Case Study: The Exxon Valdez

An examination of perhaps the most famous oil spill in history, the *Exxon Valdez* incident, illustrates the direct link between a catastrophic event and the establishment of environmental regulations. Just after midnight on March 23, 1989, the *Exxon Valdez*, an oil tanker, collided with Bligh Reef in Prince William Sound, Alaska, spilling approximately 11 million gallons (over 260,000 barrels) of crude oil.[3, 5] The spill affected approximately 1,300 miles of Alaska coastline and negatively impacted the marine ecosystem of the sound.

Intense media coverage of the *Exxon Valdez* drew attention to the need for stricter regulatory controls for oil tankers at sea. Prior to the Exxon spill, of the few regulations in place, none were designed to deal with the effects of oil spills and the funding of restoration of impacted marine environments. Essentially, the *Exxon Valdez* spill paved the way for the passage of the Oil Pollution Act (OPA 90) by the United States government in 1990.[5] Under the auspices of OPA 90, the United States Environmental Protection Agency (EPA) was given the authority to enforce spill-planning on a worst-case-scenario basis, and to administer the newly-created federal spill response system. Additionally, these regulations allowed the EPA to assess damage to the natural environment caused by oil spills and to assign financial responsibility for cleanup and restoration. Under this act, the EPA was allowed to administer the federal cleanup fund established under OPA 90.[5]

The *Exxon Valdez* spill is no longer on the top-10 list of international oil spills. However, due to the isolated Alaska location, the abundance of wildlife, and the remote inaccessible coastlines, the spill is still considered to be the largest in terms of environmental impact.[5]

REFERENCES

1. American Petroleum Institute. 1997. Environmental Guidance Document: Waste Management In Exploration and Production Operations, 2nd Edition. Washington, DC: American Petroleum Institute. 78.

2. United Kingdom Department of Trade and Industry. 2003. Environmental Legislation Applicable to the Onshore Oil Industry in England, Scotland, and Wales. www.og.dti.gov.uk/regulation/legislation/environment/ onshore_hydrocarbons/index.htm

3. National Oceanic and Atmospheric Administration (NOAA), Hazardous Materials Response and Assessment Division. 1992. Summaries of United States and Significant International Spills. Seattle, Washington: United States Department of Commerce. 224.

4. Stanislav Patin. (translation by Elena Cascio). 1999. Environmental Impact of the Offshore Oil-and-gas industry. Chapter: Conventions Regulating Environmental Impact. www.offshore-environment.com/conventions.html.

5. Kim Inho. 2002. Ten years after the enactment of the Oil Pollution Act of 1990: a success or a failure? Marine Policy, v. 26, p. 197–207.

6. Ross M. William, 1973. Oil Pollution as an International Problem: A Study of Puget Sound and the Strait of Georgia. Western Geographical Series, Volume 6. Department of Geography, University of Victoria, British Columbia. 278.

7. Fridtjof Nansen Institute. 2003. Yearbook of International Co-operation on Environment and Development. Agreements on Environment and Development: the Marine Environment. http://www.greenyearbook.org

8. International Maritime Organization. 2003. International Convention for the Prevention of Pollution from Ships, 1973 (MARPOL). www.imo.org/Conventions/mainframe.asp

9. Office for the London Convention. Convention on the Prevention of Marine Pollution by Dumping of Wastes and Other Matter, 1972, http://www.londonconvention.org/main.htm

10. United Nations. Conference on Environment and Development (UNCED), 1992. "Rio Earth Summit" www.un.org/geninfo/bp/enviro.html

6.0 WASTE MANAGEMENT STRATEGIES

6.1   Introduction

Waste management is a system of practices and controls that is primarily designed to prevent the pollution of the environment. It is also only one part of a broader system of practices and controls designed to also prevent the pollution of the air, land and water. As such, preferred waste-management strategies are those that do not create or intensify environmental problems in other media (air, land).

Waste management is not synonymous with waste treatment. Although treatment is almost always essential, it alone is rarely appropriate as a total management strategy. Many waste-treatment technologies do not, in general, render pollutants harmless. They merely transfer them into another medium, where they may retain potential for causing environmental harm. For example, air-stripping is a waste-treatment process that transforms a water-pollution problem into an air-pollution problem. Chemical precipitation transforms a water-pollution problem into a land-disposal problem. Solving one pollution problem while at the same time creating another is difficult to justify unless the pollutant becomes more amenable to being rendered innocuous, or becomes more amenable to secure disposal methods.

Unfortunately, the waste-treatment processes that are commonly used in the oil industry generate byproducts or residues, which may cause air-pollution or land-disposal problems that are not necessarily more amenable to solution. For example, waste-treatment sludge that contains toxic metals can pose acute disposal problems. Recovery of metals, which is often feasible if applied to waste, is often not feasible when applied to sludge. Land disposal of sludge is stringently regulated by the provisions of a variety of regulations because of concern for the protection of groundwater resources. To be acceptable for land disposal, sludge

68

may require substantial treatment (e.g., chemical stabilization, or solidification). Certain waste-treatment sludges may, in the near future, be prohibited altogether from land disposal.

While waste treatment is unavoidable as long as manufacturing processes generate waste, treatment in general leads unavoidably to other environmental problems. Waste treatment is both a solution and a problem. Effective waste management must recognize this paradox.

The purpose of this chapter is to present strategies that, individually or in combination, may resolve this paradox to some extent. The essence of these strategies is to reduce the amount of treatment that is needed by reducing the volume or the toxicity of waste or both at its source. By reducing the amount of treatment, the problems created by treatment will also be reduced. These strategies can be categorized as follows:

> Input Substitution
> Product Reformulation
> Production Process Redesign or Modernization
> Improved Operation and Maintenance of Production
> Recovery or Recycling Equipment

Of these, the first four directly involve changing the manufacturing process or operation. The fifth option involves the introduction of additional unit processes between the manufacturing operations and the treatment operations. The unit processes used in recovery or recycling may be similar in nature to those used for treatment, but their function is different in that they do not produce by-products which can only be disposed of. Rather, they produce products that can be either reused in the original manufacturing process or can serve as an input material in a different manufacturing process. In the rest of this chapter, examples of waste management specific to the oil industry are presented.

Although these strategies have been introduced in the context of waste treatment, they are conceptually applicable not only to the management of waste, but all waste. As such, they are presented in generic terms so as to emphasize their wider range of applicability.

Input Substitution

Input materials of production processes generally fall into two classes: those that are fabricated into end-products; and those that only serve to facilitate the fabrication process. Examples of materials of the first type include arsenic, chromium, lead, and silicon. Examples of materials of the second type include mineral acids and organic solvents. Although materials of the first type are intended to be incorporated into end-products, manufacturing processes seldom, if ever, use materials with absolute efficiency; and input materials become waste. All materials of the second type, by the nature of their use, will become manufacturing waste products.

Input materials are selected because they possess properties that are essential to the function or the fabrication of the end product. Unfortunately, what is often overlooked in the selection of materials is that materials desirable as inputs often can be very undesirable as waste. If it is kept in mind that all pollutants originate as input materials, as impurities of input materials, or as side reactants, then waste management should begin with the careful selection and control of what goes into the manufacturing process.

That which does not go into the process cannot come out as waste. Of course, eliminating all inputs in order to eliminate the waste management problem is not a sensible solution. However, substituting input materials that are environmentally undesirable as waste products with materials that are more easily managed as waste may be a feasible solution if the substitute material does not compromise the quality of the end-product. Substitution of materials that are incorporated into the final product may sometimes be difficult without at the same time investigating some of the other options that are discussed below, particularly product reformulation. Similarly, substitution of materials that are used only to facilitate the fabrication of a product may sometimes be difficult without the simultaneous consideration of production-process redesign.

Product Reformulation

The selection and control of input materials is to a great extent dependent on the type of end-products being produced. As mentioned in the previous section, a change in the specifications of the end-product may provide an opportunity to reduce the use of environmentally undesirable input materials. Product reformulation can reduce both the amount of toxic materials that are used in the fabrication of the product and the amount that is incorporated into the product. This approach is particularly attractive from a "product life-cycle" perspective: that is, when one considers not only the waste-management problems generated by the creation of the product, but also those that are generated by the disposal of the product at the end of its useful life. After all, toxic substances in products can become toxic waste when the product is discarded.

Although product reformulation is perhaps the most conceptually attractive of all waste-management strategies, it is also generally the most difficult to implement. The feasibility of this strategy depends strongly on the flexibility of a product's specifications. Since many specifications reflect customer desires or performance demands, they are difficult to change without at the same time changing the nature of the product's market.

Cooling-tower operations can result in the necessity of using several chemical additives as scale inhibitors, pH control (sulfuric acid), and as biocides. Alternative scaling inhibitors can be procured that do not require the addition of pH additives, resulting in the reduction of a chemical stock. Various biocides exist that will result in safer employee handling.

Production Process Redesign/Modernization

The objective of the production process is to transform input materials into end-products. As was discussed previously, the efficiency of this process plays an important role in determining the magnitude of the waste-management problem. The more efficient the production process, the less of those materials that are intended to be incorporated into final products will have to be handled as waste. A more efficient process also generates less waste by minimizing the use of those materials that are required only to facilitate the fabrication of the product. In addition to the benefits of minimizing the cost of managing by-products, efficiency is obviously also desirable from the perspective of maximizing the utility of input materials that have a high unit cost. Process redesign or modification can also achieve waste reduction without necessarily increasing efficiency. A redesign or change may allow the use of raw materials that are more amenable to recycling, treatment, or disposal.

An example of a production-process modification consists of methods for removing or mitigating the quantity of tank-bottom sludges in crude-oil stocktanks. Sludge reduction can be achieved by adding propellers or paddles, resulting in a continuous mixing of the crude, which keeps paraffins in suspension instead of allowing them to accumulate at the tank bottom.

6.2   Improved Operation and Maintenance

Best operating practices are procedures or institutional policies within an operation that result in a reduction in waste generation. Such practices include institutional and procedural measures, loss-prevention programs, and waste-stream segregation (EPA/530-SW-86-042).

Institutional and procedural measures are those that involve the establishment of conscious and programmatic efforts to reduce the amount or toxicity of waste. Such measures include management-instituted programs for waste reduction, environmental auditing, and employee training. All of these measures are designed to instill a "waste-reduction mentality" into the design and operation of every process. The installation of secondary containment beneath drum storage locations is an example of a housekeeping operation and maintenance practice that can prevent secondary releases into the environment.

Perhaps the most conceptually direct method for reducing the volume of waste is to reduce the amount of water used in processes. Wide-open water valves are seldom essential or efficient. To achieve a cost-effective balance, the efficiency of rinsing operations must be improved. This can be accomplished either by reducing the amount of process chemicals and waste materials that are introduced into the water, by modifying the rinse method, or by a combination of both. A simple way to reduce water usage during drilling operations is to install high-pressure, low-volume nozzles on the rig hoses, resulting in reduced quantities of water in the reserve pit.

Segregating dissimilar or incompatible waste streams does not reduce the volume of waste, but it does provide an opportunity for the recovery of contaminants and for less complex and more effective treatment. The recovery of highly valued chemicals is often more viable from more concentrated solutions, and their potential for reuse is also enhanced. Segregating waste streams, even when the opportunity for recovery or recycling is not enhanced, can still be important to the overall waste-management program. Treating segregated streams allow segregation of secondary pollutants (e.g., sludges). In situations where some of these secondary pollutants are more difficult to manage than others, segregation affords the opportunity for significantly reducing the cost or complexity of final disposal. For example, the disposal of many sludges that accumulate on slop-tank bottoms is highly regulated under the hazardous waste provisions of RCRA. These sludges could be designated as "listed" hazardous waste under 40 CFR, Part 261.

6.3   Recovery/Recycling

The objective of recovery and recycling is to take either the entire waste stream or specific constituents from the waste stream and reuse them either in the original manufacturing process or as an input material in some other process.

Waste recovery and recycling opportunities exist in both upstream and downstream operations. Examples include returning absorbent materials used to control equipment-lubrication leaks to the vendor for recycling, returning spent carbon filters from the flaring process to the vendor, and reusing spent caustics from drilling operations to reduce the acidity of drilling fluids.

7.0 TREATMENT TECHNOLOGIES

7.1 Waste-Reduction Technologies

Waste-reduction technologies are waste-management techniques adopted by various industries and manufacturers to eliminate waste disposal. The reduction approach focuses on reducing the amount and toxicity of waste by preventing it from being created—otherwise known as source reduction, which also includes recycling. These techniques can be practiced at several stages in most waste-generating processes. Various forms of waste-reduction technologies applicable to the oil-and-gas industry are discussed below.

7.1.1 Solids Control

Solids control in the oil-and-gas industry refers to the management of solids associated with the geological displacement of earth matter, as well as solids generated from manufacturing activities. The contaminants in most solids vary from industry to industry. Manufacturing processes can affect the nature of the solid waste generated. The ability to establish a good solids-control system is reflected in the overall production cost. Solids-control technologies have proven effective in the following production areas:

- Oil-and-Gas Drilling
- Distilling
- Tunneling
- Slop-Oil Treatment

- Water and Wastewater Treatment
- Sludge Dewatering and Thickening

Most drilling operations require a tremendous amount of drilling fluid. The constituents of the drilling fluid pose an environmental issue if they are disposed of as drill cuttings. To eliminate or reduce the environmental impact of drill cuttings, several procedures have been improved over the years. For instance, oil-based drilling fluids have gradually been replaced by synthetic-based drilling fluids. However, this change does not guarantee safe disposal of drill cuttings. With regard to offshore drilling, environmental pressures have resulted in the minimization of discharging synthetic cuttings into the sea. New drilling techniques that increase efficiency have raised concerns regarding zero-discharge of drill cuttings and the minimization of synthetic fluids in discharged cuttings (e.g., the use of closed-loop reserve-pit systems and improved solids control such as with a centrifuge system). Solids-control techniques such as dilution or hydrocycloning of the drilling mud are also used in oil drilling to reduce the accumulation of solids.

Other solids-control techniques that have proven effective in oil-and-gas exploration are implemented in the following processes:

- Synthetic reduction in drill cuttings
- Cuttings re-injection (CRI)
- Bulk shipment of drill cuttings (bulk-slurry transfer systems and pneumatic-bulk transfer systems)
- Skip-and-ship transfer systems (including vacuum and auger feeds)

7.1.2   Synthetic Reduction in Drill Cuttings

This process increases the efficiency of solids-removal by positioning dryers downstream of the rig solids-control system and fitting rig shakers with finer screens. The dryers consist of a vertical centrifuge that dries the cuttings by using a centrifugal force of up to 300 g. The effluent of the centrifugal dryers is processed through a high-speed centrifuge for optimal cleaning of the recovered mud. Additional mud lost over the shakers is collected in the cuttings-dryer and returned to the active mud system once the solids are centrifuged. This system reduces synthetic discharge to less than 3% by weight.

Screening improves mud-cleaning by separating out undesired fine solids before re-circulating the mud down the drill hole. However, using finer screens can result in the degradation of drill cuttings into ultra-fine material. Only two solutions currently exist for managing ultra-fine drill solids: mechanical removal with centrifuges or dilution with costly synthetic-based fluids.

An alternate technology, the squeeze press, does not depend on centrifugal force to strip the liquid, but works through a gentle squeeze-and-aerospace, vibrating-wedge wire-screen process, which squeezes the maximum amount of liquid-phase mud available. The mud-recovery objectives for this process are given in Table 7.1.

Table 7.1   Mud-Recovery Objectives

Objectives
• Dry cuttings to OOC < 3%, avg. 1.8%
• Recover/recycle all mud from cuttings
• Reduce dilution costs
• Improve cuttings-transfer systems and reduce space requirements
• Reduce haul-off volumes and disposal and treatment costs

7.1.3   Re-injection (CRI) Cutting

Re-injection of oil-contaminated drill cuttings is attracting considerable attention as a cost-effective means of complying with environmental legislation concerning discharges of oily wastes. The technique normally involves collection of waste from solids control equipment on the rig, followed by transportation to a cuttings processing station. Cuttings are milled and sheared in the presence of water, usually seawater. The resulting slurry is disposed of by pumping it into a dedicated disposal well, or through the open annulus of a previously drilled well into a fracture created at the casing shoe, which is set in a suitable formation.

The techniques utilized for disposing of waste cuttings from platform-based operations can usually be narrowed down to two choices: either re-injection into a dedicated disposal well, which if newly drilled can be re-completed as a producer or water-injector at a later date; or re-injection through the annulus of a well drilled prior to the current live well. Drilling and disposing of cuttings into the same well is possible, but to date, because of well-control concerns, this is not a preferred option with operators.

7.1.4   Bulk shipment

Bulk-shipment operations use large boxes of variable capacity (average 230 bbl)—as opposed to small 6-ton (15 bbl) skips—to collect cuttings on the rig. Typically, a standby vessel arrives on a logistics schedule to collect the bulk-cuttings containers, and the cuttings are transferred to the vessel by blowing the dry material directly from the rig. This eliminates crane-handling of skips, reduces equipment-handling requirements during containment operations, reduces potential injury hazards, and allows the rig to continue drilling during poor weather conditions.

The process of bulk containment starts with the discharge of the drill cuttings from the vibrating shakers of the rig and transferring the cuttings by auger or vacuum to a vertical centrifuge. The cuttings enter the upper cavity of the vertical centrifuge by means of gravity, flowing into the internal bowl of the typhoon, where the solids and liquids are separated by centrifugal force. During the separation phase, the bowl rotates at 230 G. A fine mesh screen is fitted into the internal bowl of the rotating vertical centrifuge in order to facilitate separation of liquids and solids.

Liquid mud recovered by centrifugal force flows from the dryer via a 4-inch conduit, to a high-speed centrifuge. Low-gravity solids are removed from the liquid phase of the mud at 3,100 G. The clean mud is transferred back to the mud-pits using a mono pump fitted onto the centrifuge unit.

Gravity causes the solids phase of the mud to fall into large containers designed to fit onto the available envelope of the rig and to have enough capacity to store the hole-sizes planned. Cuttings are usually stored at a rate of about two times the hole volume, as opposed to competitive systems reporting six to seven times the hole volume. This allows for more hole depth to be drilled and stored awaiting the arrival of the vessel. The blower system is positioned adjacent to the bulk containers on the rig for transporting the dry cuttings to the manifold on the vessel during filling operations.

Upon arrival at a port, the cuttings-transportation vessel will be secured to a bulk docking facility located near the thermal-desorption facility. The blower unit fitted on the vessel will retrieve the dry drill cuttings from the bulk containers and transport them via a flexible conduit line to the thermal facility receiving area. Some of the bulk-shipment objectives involved with this process are given in Table 7.2.

Table 7.2   Objectives of Bulk Shipment

Bulk Shipment Objectives
• Eliminate skip transfers/rentals
• Improve health and safety
• Dry cuttings to OOC < 3%, avg. 1.8%
• Store more cuttings per hole drilled
• Recover/recycle mud from cuttings
• Decrease dilution costs
• Reduce haul-off volumes and disposal and treatment costs
• Improve cuttings transfer

7.1.5   Annular Re-injection of Waste Cuttings

Annular re-injection is a method of re-injecting produced waters such as brines into the production well. Sequential annulus injection is invariably the preferred means of cuttings-disposal, particularly at offshore locations, due to its flexibility and its ability to avoid the cost of drilling a dedicated disposal well. For cost reasons, dedicated re-injection wells are usually practical only on land or in shallow water. However, their advantages include ease of cleaning out with coiled tubing in the event of plugging, possible accommodation of high volumes of waste, the ability to inject larger-sized solids, and a reduced risk of tube-plugging. Even so, unless annular cuttings re-injection is not viable—for example, because of lack of annular access to a suitable deposition horizon—drilling a dedicated disposal well is usually ruled out on cost grounds. Thus, annular re-injection of waste cuttings is invariably the method of choice. Typically, the 13 3/8" by 9 5/8" annulus is selected as the disposal location.

7.1.6   Skip-and-Ship

This system uses screw conveyors or vacuum units to transport drill cuttings from the solids-control equipment directly into cuttings containers.

Screw conveyors have been the most widely used method for transportation of drill cuttings for several years in the oil industry. Because of the screw-conveyor track record of breakdowns, high maintenance requirements, failures, and injuries resulting from exposed moving parts, they are no longer the preferred method for cuttings-conveyance. In addition, screw-conveyor installations require costly hot work and extensive labor, and may require deck or wall penetrations to route the units. The industry now considers the screw conveyor to be a potentially hazardous and costly method of conveying drilled cuttings. Currently, the focus is on using new vacuum technology to convey drill cuttings.

Vacuum equipment is completely enclosed and does not have any exposed moving parts, eliminating the possibility of injury to personnel. The installation of bulkhead penetrations does not require hot work. Vacuum-line routing is relatively simple and installs quickly. By installing bends along with horizontal and vertical runs of hose or PVC pipe, conduits are established for airflow to move cuttings from the rig-shaker collection point to their end destination, bulk containers.

The vacuum system uses a hose from the cuttings-ditch collection point to retrieve drill cuttings. As the solids and liquids travel through the hose conduit, the solids phase falls out of suspension into the container while the air phase continues through the vacuum unit and exhausts into the atmosphere. This system can process up to 25 tons per hour of drill cuttings per vacuum unit.

Four to six containers (engineered at 1 bar or 14.7 psi) are positioned in a suitable envelope during filling operations and are used as vacuum vessels. There are many suppliers that state their boxes are vacuum-rated and have been used in vacuum applications. However, investigations routinely suggest that the boxes are not properly engineered for full vacuum loading, and may be used in an unsafe manner.

It is frequently assumed that skips (conveyors) are not adequately vacuum-rated and thus require the usage of cuttings vacuum-hoppers. Typical vacuum-hoppers cannot discharge and vacuum simultaneously. Thus, two gravity vacuum-hoppers are required to handle 100% of the drill rate, with no redundancy resulting from equipment faulting. Additional problems may result from drill cuttings being sticky and not flowing freely from the gravity-hoppers, which in turn compromises the drill rate. To eliminate concerns, a system that utilizes a mechanical screw within the vessel pushes the cuttings out of the hopper. Each continuous vacuum-discharge hopper can handle 100% of the drill rate. The unit is equipped with an articulating arm, which can be positioned over the containers during filling operations. This system is capable of transporting cuttings of a 12 1/4" hole section at a rate of 100 meters per hour.

7.1.7 Design Considerations

The main problem related to oilfield slurries is their heterogeneity, which is a function of the rock material, retained fluid characteristics (such as OBM), and characteristics of the diluting fluid. Slurry viscosity is affected by a number of factors, such as particle size, shape or concentration of the particles, physical-chemical interaction between particles and liquid (e.g., flocculation, hydrodynamic interaction), and liquid viscosity. Waste-slurry particle characteristics will depend on the lithology drilled as well as the drilling parameters for the given well profile (bit type, mud type, solids-control equipment, WOB, trajectory, etc). With such a list of variables to consider, it is safe to assume that no two slurries will be the same, and that engineering solutions for slurry-handling must be tailored on a case-by-case basis. Particle cohesion due to the grinding effect of the pumps used to circulate the slurry is also an issue. This process generates increasingly smaller particles, thereby exposing new surfaces to be wetted by the liquid(s) present. Large increases in slurry viscosity can be expected during this process, which could potentially limit the ability of the pumps and the circulating system to handle the slurry.

Engineers can control these slurries by considering the particle-particle and particle-liquid interactions. By using treatment chemicals it should be possible to engineer slurry viscosity to achieve three objectives:

1) Optimize the solids-loading of the slurry (i.e., minimize dilution)
2) Optimize slurry viscosity to enable pumping and slurry transfer
3) Optimize slurry stability to minimize settling during transit

Achieving these objectives requires a technique to allow the measurement of slurry viscosity during the optimization process.

The volume of excess drilling-fluid generated while drilling is dependent on the volume of dilution required to maintain desired drilling-fluid properties. The amount of dilution required depends upon the characteristics of the formations being drilled, the concentration of drilled solids tolerated in the mud, and the effectiveness of the solids-removal process.

7.1.8   Drilled-Solids Concentration and Dilution

The volume of dilution required to compensate for the incorporation of a barrel of drilled solids is given by the following equation:

$$bbls = (100 - \%)/\%$$

Where: bbls = dilution required/bbl of incorporated solids, and % = desired (tolerated) drilled solids concentration in percent volume.

Good mud-engineering practices require that concentrations of drilled solids are limited to low levels—generally at no more than 5% but occasionally as high as 8% by volume, with fluids weighing less than about 12.0 pounds per gallon. At higher densities, the concentrations must be reduced to a typical range of 3% to 5%. Using the equation, at 5% by volume drilled solids, each barrel of incorporated solids requires the preparation of 19 barrels of new mud to maintain the limiting drilled-solids concentration. Conversely, the removal of one barrel of drilled solids from the system eliminates the need for the preparation of this amount of fluid. While this is only approximate because of down-hole fluid losses, there is usually little error, considering that all of the prepared mud eventually becomes excess material that must be disposed of. The total cost of dilution, then, is the sum of the cost of preparing the fresh mud used for dilution, plus the cost of disposing of an approximately equal amount of fluid.

Even at 8% drilled solids by volume—a concentration too high to be recommended with water-based fluids—the incorporation of one barrel of drilled solids requires the preparation of 11.5 barrels of new mud. These dilution figures clearly illustrate that solids-control effectiveness has a very significant impact on both drilling-fluid and drilling-waste disposal costs.

In weighted drilling fluids, the most common and troublesome contaminant is excessive concentrations of ultra-fine and colloidal solids, both barite and drilled solids. Barite, similar to many drilled solids, is relatively soft. Consequently, particle size diminishes with use, thereby increasing the concentration of ultra-fine and colloidal particles, and creating rheological problems. Viscosity increases as the average particle size decreases.

The centrifuge is the preferred tool for the control of this problem. There are only two options: reducing the concentration of undesirable fine solids by dilution, which requires the disposal of large volumes of excess fluid, and the subsequent loss of desirable barite; or removal with the centrifuge. Even though centrifuging and discarding the centrate (overflow) entails the loss of most of the liquid and dissolved chemicals in the processed fluid, it permits the retention of the desirable-sized barite and greatly reduces the need for dilution as a means of viscosity-control, thereby reducing the volume of waste generated by drilling.

7.1.9   Operational Considerations and Planning

A variety of operational details must be considered in order to properly plan a drilling project. Successful operations dictate that the majority of the work must be completed in the planning stages. Some of the details include:

- Identifying suitable cuttings disposal/sealing formations
- Selecting surface equipment
- Designing the casing program
- Designing the injection program/contingency planning
- Plug prevention in the annulus and the formation
- Preventing cuttings slurries from breaching to the surface or contaminating the water table
- The impact on existing producing wells or future wells
- Quality control/monitoring of injection procedures
- Permanently entombing abandoned waste
- Obtaining regulatory approval
- Addressing environmental and safety concerns

Each of these operational considerations is directly affected by the characteristics of the subsurface environment, sealing formations, injection zones, slurry properties, drilling plans, subsurface slurry disposal dimensions, and other elements. Of the various technical questions that must be evaluated, the least-understood (but nonetheless important) are those associated with down-hole considerations:

- Into what formation can the cuttings slurry be injected?
- How will the cuttings slurry be contained?
- In what direction will the cuttings slurry propagate? And how far?
- How significant an impact will the cuttings slurry have on nearby well bores/formations?
- How will the cuttings slurry affect existing wells and future drilling plans?
- What volume of cuttings slurry can be safely disposed of?
- What forces will be put on the well casing?
- How are slurries injected in order to minimize formation impact?
- How are the annulus and the formation protected?
- When the formation changes, how does the CRI operator vary the slurry specifications?
- Is particle size < 100 microns?

Proven equipment reliability, manpower requirements, utilities, ease of installation, and contingency plans must all be considered when designing the surface equipment system. Proper system design is crucial, since any downtime for repairs or maintenance directly affects drilling progress. In zero-discharge operations, it is important that the CRI surface equipment is adequately sized to process the cuttings ahead of the drill rate/surge conditions. Contingency plans should be developed for each specific project.

7.1.10 Results of Case Study

7.1.10.1 Waste Drilling-Fluid in Conventional Earthen Pits

A small independent operator was concerned about the volume of drilling waste in conventional reserve pits at his drilling locations. Waste-management costs were a concern, as well as the costs associated with the impact of the drilling fluid on the adjacent land due to earthen reserve pit failures. The operator was concerned about the potential for surface-water or ground-water contamination and the associated potential liabilities.

As a solution, the operator implemented a procedure change using a closed-loop drilling-fluid system rather than the conventional earthen reserve pits. The wells were to be drilled in relatively shallow, normally pressured strata, and therefore were amenable to the usage of a closed-loop system. The operator negotiated with drilling contractors to obtain a turnkey contract that required the drilling company to use a closed-loop system and take responsibility for recycling the waste drilling-fluid generated by the operation. The turnkey contract was incrementally more expensive. However, the operator realized that there would be savings of about $10,000 per well because of reduced drill-site construction and closure costs, reduced waste-management costs, and reduced surface-damage payments. Most importantly, the drilling mud was reclaimed and reused rather than disposed of by burial or land application. Also, the operator reduced the potential for environmental impact and associated potential liability concerns.

7.2 Landfilling/Land Disposal

7.2.1 Applicability of Technology

Landfilling, or land disposal, is one of the oldest and most popular methods of managing anthropogenic waste materials in which contaminated soil, sludge, or other media are systematically placed in a large excavation at the site of the operation. Accumulated waste is compacted and packed in a specific manner on a daily basis. Once a landfill has been completely filled, it is capped with an impermeable layer and monitored for leachate (materials that collect at the bottom of a landfill over time due to seepage) and methane.

Landfilling can be useful for the disposal of a multitude of contaminated substances. The process is limited by regulations regarding the use of landfills as established by national or international environmental agencies around the world. Biodegradation is the only means of achieving contaminant removal within the landfill. Nutrient addition and other methods of soil modification can be used to stimulate biodegradation. Composting and the use of biopiles are two similar waste-management operations that involve greater efforts at stimulating the bio-degradation of contaminants, as compared to landfilling.

7.2.2 Description

The basic procedure for operating a landfill is usually identical for all applications. Some applications, however, may be more sophisticated with regard to the steps that are followed for waste placement. The landfilling process consists of site assessment and preparation, landfilling or waste-placement activities themselves, site closure operations, and monitoring activities. The stages of a landfilling operation are described in Figure 7.1.

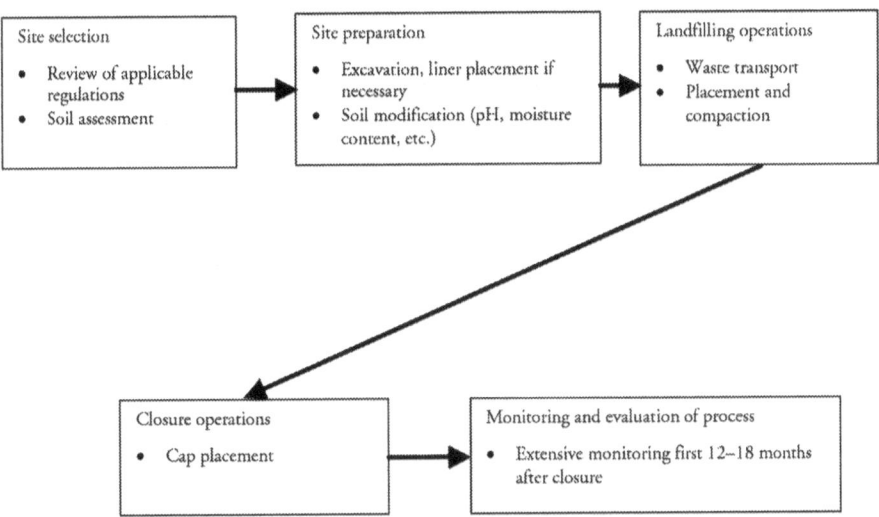

Figure 7.1   Landfilling/Land Disposal Process Flow Chart

Several tasks must be performed at a potential landfill site before the waste is landfilled. Designers must decide whether or not an excavation will be required. Waste is preferably deposited below grade, necessitating a pit or excavation. Often, designers will attempt to use natural site topography to minimize the potential costs associated with excavation. Once a site has been identified, relevant government and state regulations must be consulted to ensure that landfilling is performed lawfully. The EPA sets the majority of guidelines for the operation of a landfill in the United States. Jurisdictional agency regulations or legislation may apply also, specifically "land bans" that prohibit the application of certain wastes to land in a particular area or region. Following a careful review of applicable regulations, assessments of the soil on site are performed. The goal of soil assessment is to determine the suitability of the land for proposed structural and potential environmental impacts. Characteristics of the soil that are investigated include pH, moisture content, humic content, geotechnical parameters, and microorganism population. If soil characteristics are discovered to be suitable, site preparation can begin.

Site preparation generally consists of liner placement and as-needed soil modification. Whether or not a liner will be employed is largely dependent on the soil characteristics of the site, and government regulations. The soil's capacity for biodegradation and susceptibility to infiltration are the main parameters taken into account when making this decision. One of three liner-types is generally employed: a clay liner, a plastic liner, or a composite liner constructed of plastic and compacted soil.

Waste-placement activities follow landfill site preparation. The waste to be landfilled is collected and transported to the site where the waste-deposition process ensues. Non-compatible wastes are separated into individual control cells that are organized into two- to three-foot layers.[1] The wastes are subsequently compacted and covered with soil. The covering soil is often taken from a stockpile established prior to waste deposition. The soil must control odor and prevent rainwater infiltration into the landfill. Waste deposition continues until the cell in use is full.

Once all cells of the landfill have been filled, they are capped and closed according to regulations. When the entire landfill has been used to capacity, it undergoes closure operations, which consist of placing cap materials over the landfill. The cap is typically made of a low-permeability, low-maintenance material that provides a durable, infiltration-proof cover for the landfill.

For the first 12 to 18 months after landfill closure (or as determined by jurisdiction regulations), extensive monitoring activities are performed at the site. The site is routinely inspected and local media are sampled to determine whether waste has contaminated soil or groundwater. If no problems are detected within the first 18 months of monitoring, the frequency of site inspections is gradually decreased. However, monitoring of leachate and methane is an ongoing process at landfills. The leachate is either collected and disposed of as a waste material or is re-circulated back into the landfill to expedite the decomposition of the waste. Methane gas is vented, collected and flared, or, in some instances, collected and used as a source of energy.

7.2.3 Advantages and Disadvantages of Landfilling/Land Disposal

Landfilling has several characteristics that are considered advantageous for soil remediation applications. A broad range of contaminants can be landfilled. A single landfill site can be utilized to treat a very large volume of contaminated media relative to the treatment capacities of alternative soil-remediation technologies. Additionally, landfilling is neither a technically nor scientifically complex process, therefore making it a relatively easy treatment technology to implement.

A number of disadvantages are also associated with landfilling. It has been argued that the merits of landfilling do not justify potentially devastating effects to local environments and ecosystems. Low public opinion of landfilling has developed as a result of these debates. Furthermore, landfilling is generally not considered a method of final treatment, but a method of safely storing the waste. Landfills typically must be operated with sporadic monitoring for many years. In addition, there are significant operating-space requirements for landfill use, along with numerous regulations that must be followed and permits that must be acquired.

Landfill operators must be cognizant of any land bans that may be applicable to media that is to be treated. Local or federal regulations can prohibit the deposition of certain contaminated substances at specific locations. Disposal of liquid waste at landfills is also an inappropriate practice. Moreover, there is potential for contamination of groundwater underlying a landfill, despite efforts to decrease permeability at the base of the landfill.

7.2.4 Design Considerations

The design of a landfill site involves the consideration of several key components. Adherence to local, state, and federal regulations for landfill operation is of utmost importance. Particularly, the EPA has developed detailed guidelines for landfill operations. As previously mentioned, several wastes have been deemed inappropriate for treatment by landfilling. EPA regulations also stipulate that various monitoring requirements be met for lawful landfill operation. Subsurface ground-water, the vadose zone, and gas emissions must all be diligently monitored.

Leachate is another persistent concern associated with landfilling. The primary function of the landfill liner is to prevent the leaching of toxic contaminants into local soil. The cap also reduces precipitation infiltration, helping to decrease leach-ate volume. Leachate is funneled from the ends of the landfill to a central sump that removes the leachate for treatment.

Gas emissions must be taken into consideration during landfill planning and operation. Gas is produced as a byproduct of the anaerobic decomposition of organic matter in contaminated sediment. Wells are used to measure concentra-tions of methane and carbon dioxide, which are the primary gas emissions of concern. Methane is of particular concern, as it can be flammable or explosive in high concentrations and is malodorous in the presence of other gases. Gas emissions are typically vented to the atmosphere or flared. Some landfills employ energy-recovery operations that utilize gas byproducts to power other landfilling process components.

Surface water and runoff at the landfill must be controlled. Control structures are designed and constructed to funnel runoff away from landfilled media. In the event that runoff does come into contact with landfilled media, the water is sampled to determine the extent of contamination. Contaminated runoff is generally treated on-site or disposed of off-site, with treatment methods such as siltation basins remaining optional.

Several site characteristics are also considered during the design process. Weather is important to consider, as it will affect soil-excavation procedures and may complicate the storage of useful cover-soil on site. Groundwater character-istics must be determined prior to landfill operation. The location and depth of local aquifers as well as their flow-directions will influence landfill orientation. Additionally, project administration and the construction of on-site control struc-tures are affected by site characteristics. Land suitable for the construction of buildings and access roads must be available in the vicinity of the landfill for a site to be useful.

7.2.5 Environmental Effects

The effects of landfilling on the environment have been well documented. Perhaps the most significant effect of landfilling on the environment is groundwater contamination. Leaks from containment devices employed by land-fills are frequent, often resulting in the contamination of local subsurface media. Leaching of landfill contaminants to local media is one of the most complicated issues associated with landfilling. Leachate collection, surface-water management, and control-structure maintenance are all vital components of landfill operation, functioning to mitigate undesirable environmental effects.

Gas emissions from landfills can also be of detriment to the environment. Methane and carbon dioxide can be produced in large quantities at landfills. Venting or flaring is frequently used to manage gas emissions. Also, landfill material is very often malodorous, and public complaints can result from local residents' objections to a foul landfill odor.

7.2.6 Engineering Economics

The following is a list of fixed-cost and variable-cost items required for establishing a landfill operation.

- Fixed-Cost Items
  - ○ Land acquisition
  - ○ Leachate-collection system installation
  - ○ Project-facilities construction
  - ○ Permit acquisition
  - ○ Soil excavation

- Variable-Cost Items
  - ○ Operating and maintenance labor
  - ○ Equipment fuel
  - ○ Site supervision
  - ○ Site quality assurance and health-and-safety support
  - ○ Sampling and analysis for process control
  - ○ Utilities
  - ○ Post-closure maintenance and management

Landfilling is generally regarded as a relatively inexpensive means of managing solid waste.

7.3 Landfarming

7.3.1 Applicability of Technology

Landfarming is an innovative commercial treatment technology that uses bio-degradation to reduce or eliminate existing contamination in solid media. Contaminated solids are applied to lined beds where frequent tilling is employed to aerate the waste, subsequently stimulating the biodegradation of contaminants in the sediment. The treatment is an *ex-situ* process, occurring with the soil having been removed from its original location. Landfarming is also known as solid-phase biodegradation, land application, land tilling, and land treatment. The technology has been used to treat a variety of contaminants, with varying degrees of success. The types of contaminants that are amenable to landfarming are shown in Table 7.3.

Table 7.3   Landfarming Treatment Contaminant Reduction List

Contaminants Susceptible to Landfarming Treatment
• Diesel fuel
• Fuel oils, including No. 2 and No. 6
• JP5 (jet fuel)
• Oily sludge
• Wood-preserving wastes—creosote, pentachlorophenol (PCP), and polyaromatic hydrocarbons (PAHs)
• Coal residues (coke wastes)
• Various pesticides

7.3.2 Description

Landfarming has several characteristic stages and process components. Depending on the nature of the site and target contaminants, the difficulty of implementation is often variable. Typically, the landfarming process includes site preparation, soil pretreatment, treatment initiation, and byproduct management.

Site preparation involves site selection and modification to fit the planned landfarming treatment scheme. Modification of the site often entails grubbing, clearing, and grading of the existing land. Berms are constructed to prevent run-on and run-off in the area of the landfarming operation. A liner, if selected for use in the application, is also constructed during this stage.

After the site has been prepared, soil pretreatment activities can commence. Controlling the condition of the ambient soil optimizes the rate of biodegradation. Soil moisture content and pH are modified as necessary, and other amendments, such as soil-bulking agents and nutrients, may be added. Large pieces of debris with diameters in excess of 60 millimeters are usually removed from the soil as well. Aeration or tilling frequency is set once the soil conditions are determined.

Treatment is initiated following site preparation and soil pretreatment. Contaminated soil is spread over the land and lifts of variable size are used for holding and tilling. Biodegradation proceeds as a result of the continuous interactions between contaminants, soil, climate, and microorganisms. In many cases, contaminated soil is added to biodegrading soil in order to expose the contaminated soil to the microbial culture. Biodegradation is often a slow process, particularly for large volumes of contaminated soil. Despite the relatively slow speed of the process, landfarming generally results in a substantial reduction or elimination of target contaminants.

Landfarming creates several pathways for release of harmful pollutants into the environment. Byproducts and exposed toxic materials must be diligently managed. Leaching of contaminants into the soil and groundwater is a persistent concern. Liners are the primary means of deterring leaching at landfarming sites. Hazardous vapors are also sometimes generated during landfarming operations. Gas emissions from volatile contaminants must be eliminated or controlled. Greenhouses, plastic tunnels, or plastic cover-sheets can be utilized to capture most of the harmful vapors. Figure 7.2 is a schematic diagram of the landfarming process.

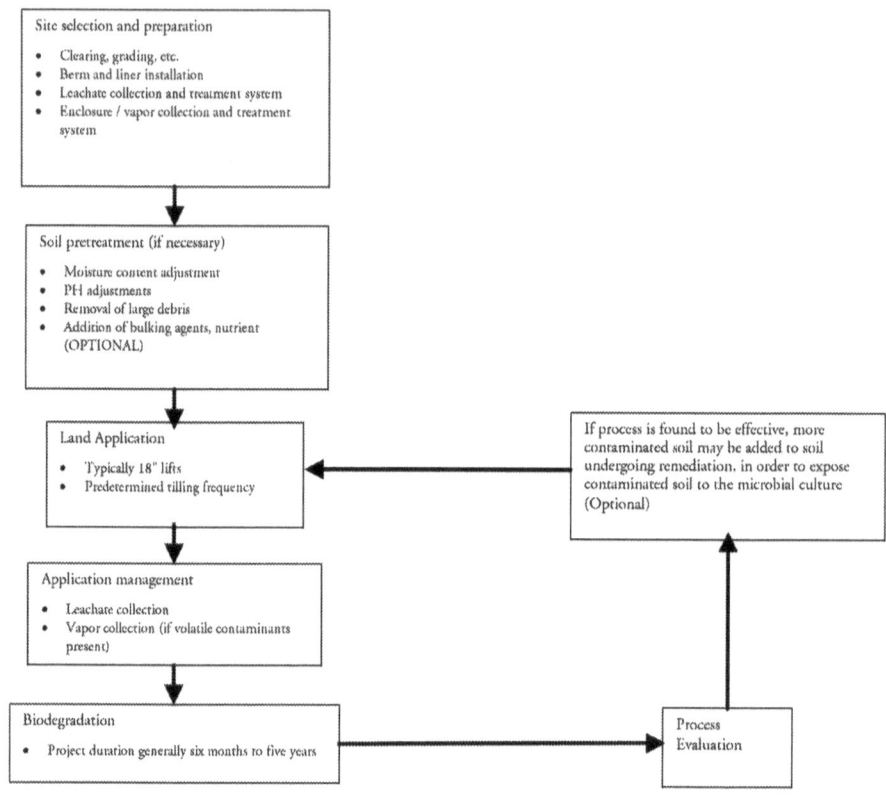

Figure 7.2   Landfarming Process Flow Chart

The duration of a landfarming application is highly variable and can range from six months to five years. Project duration is influenced by numerous factors, including remediation objectives, climate conditions, achievable biodegradation rates, and contaminant concentrations.

Advantages and Disadvantages of Landfarming

There are several advantages associated with the use of landfarming for site remediation. Contamination of non-halogenated volatile organic compounds and semi-volatile organic compounds is effectively reduced or eliminated by landfarming. The technology is also a very effective means of treating petroleum hydrocarbons.

The majority of landfarming applications are low cost. In some cases—specifically, when dealing with shallow soils less than three feet deep—excavation can be completely avoided, thereby eliminating costs associated with that particular activity. Furthermore, no residuals are produced during the landfarming process, making costly management or treatment of hazardous residuals unnecessary. The use of naturally occurring microbes for treatment can also cut costs associated with specific microbe acquisition or extensive tests to determine which microbes can effectively neutralize the target contaminant.

Although landfarming has been proven a useful remediation tactic, it has a significant number of limitations and disadvantages. The essential treatment mechanism, biodegradation, is affected by many factors beyond human control. Climate conditions such as temperature, rainfall, and wind speed have significant influences on biodegradation rates. Moreover, achieving contaminant-concentration reductions greater than 95% can be extremely difficult. Microorganisms can also be destroyed by high concentrations of contaminants in untreated soil.

The issues associated with byproduct and waste-material management can be burdensome as well. In addition to berm construction, other run on and runoff management measures, such as collection-and-treatment-facility construction, may be necessary. Leaching of contaminants into soil and groundwater at the site can be a concern. Additionally, extensive pretreatment may be required to prevent air pollution from volatile organic contaminants. The use of landfarming can also be precluded by land bans, which prohibit the application of specific contaminants to soil in certain areas.

Some other disadvantages of landfarming include but are not limited to:

- Large space requirements
- Ineffectiveness in treating inorganics
- Limited treatment depth
- Dust-control issues

7.3.3 Design Considerations

Prior to the implementation of landfarming at a particular site, many important factors are given careful consideration. The characteristics of the contaminated soil are of great importance. Landfarming is most efficiently implemented under specific soil conditions. In particular, well-drained soils are often most appropriate for the technology. A pH of 6 to 8 is generally acceptable, with a pH of 7 being optimal. Lime can be added to the soil to increase its pH, while sulfur can be added to lower it. The moisture content of the soil must also be noted and should be between 12% and 30% (40–85% field capacity/water holding).[1]

The characteristics of the indigenous microbial population in the soil to be landfarmed must be determined. Nutrient content must be controlled to preserve an ideal habitat for the microorganisms. Aerobic microorganisms such as bacteria, algae, protozoa, and fungi may all be present in the soil. Bacteria are generally most abundant and require carbon, nitrogen, and phosphorus for basic life functions. Cow or chicken manure can be added to the soil on site to provide additional nutrients.[1] Sulfate and phosphate must be monitored, as they can impair microbial metabolisms. Additional soil characteristics that are given consideration prior to landfarming implementation are permeability, topography, vegetative cover, and subsurface geological or hydrogeological features.[2]

The contaminant concentration in the soil to be treated is of equal importance during preliminary design. Depth and distribution profiles of the contaminants are essential. The presence of certain constituents, particularly chlorine and nitrogen, can complicate a landfarming application. The concentrations of organic and inorganic compounds are important parameters. Landfarming technology is more effective if the contamination is within a reasonable range. For hydrocarbons, a concentration of 50,000 parts per million (ppm) or less is usually treatable. Contamination involving heavy metals at concentrations of 25,000 ppm or less is suitable for treatment.[3] Certain types of contaminants are better suited for the process, especially those with low molecular weights. If the process involves dealing with contaminants that have high molecular weights, slower degradation rates should be expected.

Climate concerns are also prevalent when planning a landfarming application. Landfarming is a process best suited for warm climates, where it can be performed year-round. In regions where cold weather frequently occurs, the landfarming season will be shorter, usually between seven and nine months. Cold-weather applications benefit from the installation of an enclosing structure and the addition of psychrophiles, bacteria that thrive in cold temperatures. Run on, runoff, and erosion are additional climate-related issues that need to be addressed. Berms are the primary means of preventing run on and runoff at the site. Erosion can be prevented by terracing soil to the windrows or by spraying the soil to minimize dust. Generally, precipitation in excess of 30 inches per year at a site warrants the use of a precipitation deterrent. Tarps, plastic tunnels, and greenhouses can all be used to prevent rain and wind from disturbing the soil.[2]

7.3.4 Environmental Effects

Landfarming can have a multitude of effects on the environment. Volatile contaminants undergoing treatment are a potential source of air pollution. The weather conditions in the area can create contaminated runoff that migrates to adjacent lands. Similarly, contaminants can migrate to the groundwater table through leaching.

The effects of landfarming on animals, wildlife, and microorganisms have been extensively investigated. The main exposure pathway for animals and wildlife near a landfarming operation is ingestion. Plant materials, animal feed grown in the soil, or even the soil itself could be consumed or contacted by indigenous wildlife. Despite the contamination in the soil, the potential toxic effects on animals and wildlife are minimal. Assessments of the effects of landfarming on local microbial populations have been less conclusive, as the presence of landfarming operations has revealed both beneficial and detrimental effects on microorganisms.

7.3.5   Results of Case Studies

7.3.5.1   Scott Lumber Company Superfund Site

Landfarming technology was used in the full-scale cleanup of the Scott Lumber Company Superfund Site in Alton, Missouri. The contamination at this site was the result of wood-treating operations for the preservation of railroad ties, performed between 1973 and 1985. A creosote-diesel fuel mixture was used in the process, creating mainly PAH and benzopyrene contamination. This remediation project was one of the first landfarming applications involving creosote.

Clay liners, berms, run-on swales, and a subsurface drainage system were all constructed as part of the landfarming system. A retention pond and an irrigation system were also installed. Two lifts of soil were utilized, with nutrients added only to one of the lifts (designated as Lift 1). Previous studies had indicated that the indigenous microorganism population would support biodegradation for this landfarming application.

Approximately 15,961 tons of soil was targeted for treatment. Benzopyrene concentrations in the soil ranged from 16 to 23 mg/kg at the start of the project. Concentrations of PAHs varied depending on the contaminated medium in question, and were as high as 0.326 mg/kg in lagoon water, as high as 12,400 mg/kg in sludges, and as high as 63,000 mg/kg in soils. Remediation goals were set for a total of 16 PAH constituents and benzopyrene. PAHs were to be treated to a level of 500 mg/kg, and benzopyrene was to be treated to a level of 14 mg/kg.

The cleanup project lasted from December 1989 to September 1991. Remediation goals were met for all target contaminants at the Scott Lumber Company site. After six months in Lift 1, PAHs had been reduced from 560 mg/kg to 130 mg/kg, and benzopyrene was reduced from 16 mg/kg to 8 mg/kg. Following three months of treatment in Lift 2, PAHs were reduced from 700 mg/kg to 155 mg/kg, and benzopyrene was reduced from 23 mg/kg to

10 mg/kg. The total cost of the landfarming operation was $4.047 million. This reflects land contractor costs for 3 years at $1.292 million per year, plus $254,000 for lab analysis, contractors, and oversight.

7.3.5.2   Bonneville Power Administration—Ross Complex

Beginning in 1939, wood-preserving operations at the Bonneville Power Administration Ross Complex in Vancouver, Washington, caused numerous spills of waste at the site. High-molecular-weight polycyclic aromatic hydrocarbons (HPAHs), along with pentachlorophenol (PCP), were the major constituents of the contamination. A full-scale cleanup effort involving landfarming technology was initiated at the site in November of 1994.

A four-bed landfarming system was designed for use in this application. Each bed was designated as a different series utilizing a different combination of enhancement and bioremediation tactics. The combination of enhancement and bioremediation made this a significant application of landfarming technology. Lift depths of 6 to 12 inches were used, and the beds were changed every 84 days.

Approximately 2,300 cubic yards of soil was targeted for treatment. HPAH was detected in the soil at levels as high as 1,500 mg/kg, and PCP was detected at levels as high as 500 mg/kg. Three tiers of cleanup goals were set for this particular application, due to concerns over the feasibility of accomplishing initial primary cleanup goals. The primary cleanup goals, also designated as the Tier 1 cleanup goals, were 1 mg/kg for HPAH and 8 mg/kg for PCP. Tier 2 and Tier 3 goals were both set at 23 mg/kg for HPAH and 126 mg/kg for PCP. Tier 2 and Tier 3 goals differed in the type of cap that was to be put over the soil for cleanup purposes, with Tier 2 having the less sophisticated capping mechanism.

The landfarming operation at Bonneville Power lasted from November 1994 to January 1996. Tier 2 cleanup goals were met at a minimum for all contaminants. An overall reduction of approximately 80% was achieved for both HPAH and PCP. Post-treatment concentrations for the four treatment series ranged from 6.76 to 21.83 mg/kg for HPAH, and 6.8 to 20.7 mg/kg for PCP. The total cost of the operation through November 1995 was $1,082,859. This cost equates to approximately $470 per cubic foot of treated soil.

7.3.6   Engineering Economics

Some of the fixed and variable costs for landfarming operations are detailed below.

- Fixed-Cost Items
  - ○ Land acquisition
  - ○ Tilling and fertilizing equipment
  - ○ Leachate-collection system installation
  - ○ Soil excavation

- Variable-Cost Items
  - ○ Operating and maintenance, labor
  - ○ Irrigation water
  - ○ Nutrients
  - ○ Site supervision
  - ○ Site quality assurance and health-and-safety support
  - ○ Sampling and analysis for process control

Landfarming is widely regarded as an inexpensive means of treating contaminated soil. The cost of an application is largely dependent on the level of maintenance required for the operation, and land-acquisition costs.

A typical landfarming application may require a timeframe of six months to five years, during which several different costs may be incurred. Preliminary laboratory feasibility studies may range from $25,000 to $50,000, followed by a pilot test which could cost as much as $100,000.[4] Once initial tests have been performed, landfarming operators can expect to pay approximately $40 to $80 per cubic yard of soil treated with minimal leachate control. If an application requires maximum leachate control, the price rises to between $135 and $270 per cubic yard of contaminated soil treated.[5]

7.4 Thermal Desorption

7.4.1 Applicability of Technology

Thermal desorption is an innovative treatment technology by which contaminants are volatilized and separated from soil or sludge. Treatment is conducted *ex situ*, with the contaminated medium removed from its original location. The process specifically aims to volatilize pollutants, as opposed to oxidizing or incinerating them. Once volatilization is achieved, subsequent treatment measures are performed on the vapor byproduct, or "off-gas," of the initial phase of treatment.

Thermal desorption technology has been proven effective for a variety of applications treating several different types of pollutants, as shown in Table 7.4.

Table 7.4   Contaminants Susceptible to Thermal Desorption Treatment

List of Contaminants
• Halogenated and non-halogenated volatile organic compounds (VOCs)
o Gasoline, JP4 (jet fuel), other conventional fuels
• Halogenated and non-halogenated semi-volatile organic compounds (SVOCs)
o Diesel, JP5 (jet fuel), other heavy fuels
• Benzene, toluene, ethyl benzene, and xylene (BTEX)
• Polyaromatic hydrocarbons (PAHs)
• Some pesticides and polychlorinated biphenyls (PCBs)

7.4.2 Description

Thermal-desorption applications can have several different process components, depending on site and contaminant characteristics. Generally, the thermal-desorption process has three main components: pretreatment and materials-handling, desorption-unit operations, and off-gas post-treatment system operations[1] (see Figure 7.3, Thermal Desorption Process Flow Chart).

The initial stage of the thermal-desorption process involves a pretreatment and materials-handling system. Large particulate matter that could hinder treatment is removed from the contaminated soil. If necessary, adjustments to the *in-situ* moisture content of the contaminated soil are also made at this stage of the treatment process. Moisture content is modified by drying the soil or blending it with sand to achieve an optimum level for the thermal desorption process.

Pretreatment of contaminated soil is followed by operations conducted in the desorption unit. The basic function of the desorption unit is to heat contaminated soil in order to achieve vaporization/volatilization of target contaminants. There are several variations of this process, but the two main variables are: how the contaminated media contacts the heat source, and at what temperature the process is conducted.

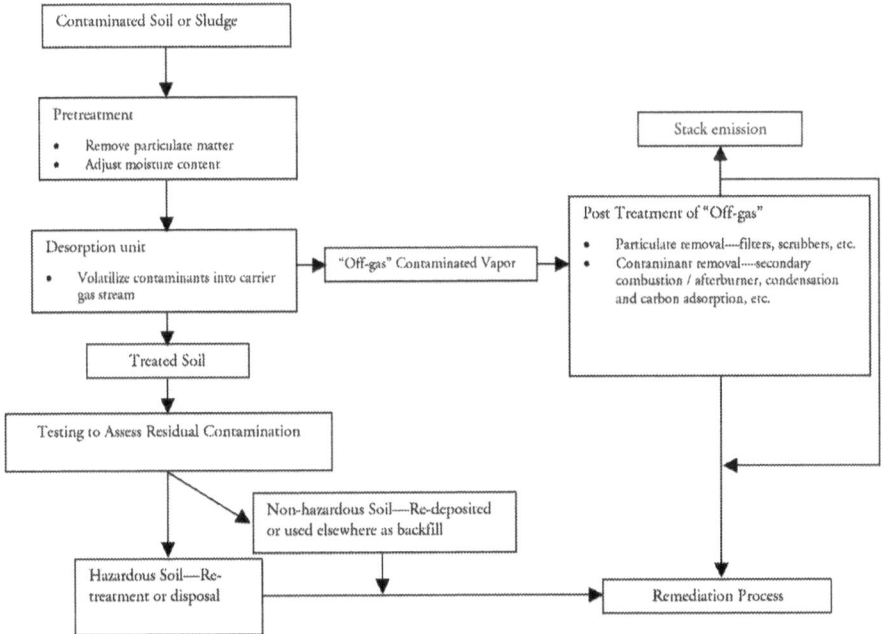

Figure 7.3   Thermal Desorption Process Flow Chart

The three common types of desorption utilized in the unit are direct-fired, indirect-fired, and indirect-heated desorption. Direct-fired desorption involves applying fire directly to the surface of contaminated media to volatilize contaminants. Some contaminants may be thermally oxidized as a result of this process. Indirect-fired desorption involves heating contaminated media within a direct-fired rotary dryer. Contact with the dryer surface results in the volatilization of water and organic contaminants. Indirect-heated desorption is performed by externally firing a rotary dryer to volatilize water and organics from contaminated media. Vapor is then removed via a carrier gas—typically an inert gas—which is selected to prevent contaminants from igniting in the unit.[2]

The temperature at which the process is conducted must also be considered. Thermal desorption has frequently been divided into two classes based on temperatures achieved during the process. Low-temperature thermal desorption (LTTD) is conducted at temperatures ranging from 200°F to 600°F (90°C to 320°C). This type of thermal desorption is frequently used for the remediation of fuels in soils. Other target contaminants are mainly non-halogenated VOCs. To destroy organics, contaminated soil must be heated to the high end of the LTTD range. If the process affects the organic components of the soil, the capacity of the soil to support any biological activity will be altered.[3]

High-temperature thermal desorption (HTTD) is conducted at temperatures ranging from 600°F to 1000°F (320°C to 560°C). This type of desorption is often used in conjunction with other remedial tactics such as incineration or dechlorination, depending on the site. Target contaminants for HTTD include SVOCs, PAHs, some PCBs, and pesticides. Certain types of volatile metals—specifically mercury—are more effectively removed using HTTD as opposed to LTTD. HTTD can be used for the removal of VOCs and fuels; however, this results in a less cost-effective treatment operation.

The post-treatment system in a thermal desorption unit treats the off-gas generated during the desorption process. After volatilization, contaminants and particulates are present as a vapor within the thermal desorption unit. A carrier gas transports this vapor to the post-treatment system, where a variety of different treatment mechanisms can be employed. Filters, scrubbers, or baghouses are devices used for particulate removal at this stage. Removal of volatilized contaminants can also be achieved by several means. Typical mechanisms or processes used for this purpose include condensation, followed by carbon adsorption, secondary combustion chambers or afterburners, or catalytic oxidizers. Treated gas is subsequently released to the atmosphere via stack emissions.

Treatment operations are evaluated following the completion of thermal-desorption treatment activities. Thermal-desorption-system performance is typically measured by comparing contaminant levels in treated soil versus those in untreated soil. Once an assessment of contaminant levels is complete, soil deemed non-hazardous can be redeposited on-site or taken elsewhere for use as backfill. Soil that is deemed hazardous may be subject to retreatment or disposal.

7.4.3   Advantages of Thermal Desorption

Thermal desorption is an attractive treatment option, as it has many advantages over other remediation technologies. A variety of different contaminants can be treated by thermal desorption, such as VOCs, SVOCs, PAHs, BTEX, some pesticides, and PCBs. The technology is also capable of separating organics from several different types of waste including refining waste, coal-tar waste, wood-treatment waste, and paint waste.[4] Some PCBs and pesticides have shown susceptibility to thermal-desorption treatment; however, incineration is the preferred method of treatment for these contaminants.

Thermal desorption is an exceptionally efficient process. The average time for cleanup of a "standard" 20,000-ton site using HTTD is just over 4 months. A single thermal desorption unit has the capacity to treat, on average, 15 to 20 tons of contaminated sandy soil per hour. Rates far in excess of this average have been regularly achieved using multiple units or advanced thermal-desorption technology. When dealing with clay soil, approximately seven tons or less can be treated per hour. These rates are highly variable and depend on site characteristics as well as target contaminants.

Another advantage of thermal desorption is its cost effectiveness. A plant conducting thermal-desorption treatment operations at a rate of approximately 18 to 35 tons per hour can expect to pay $29 to $45 per ton in treatment costs.[5] Moreover, thermal-desorption systems do not require as much fuel as other systems. In many cases, the mobility of thermal desorption units can also facilitate process implementation. Most units are mobile and can be transported directly to the site. This eliminates the costs associated with transporting contaminated waste to a treatment facility.

Other advantages of thermal desorption:

- Sand and gravel are relatively easy to treat
- Treatment systems are generally mobile
- Numerous types of thermal-desorption systems with highly variable process configurations exist
- pH adjustments on contaminated media are rarely required for process implementation
- Acid gases such as hydrogen chloride and sulfur dioxide are typically not process byproducts

7.4.4   Disadvantages of Thermal Desorption

As with most remediation technologies, thermal desorption has limitations and disadvantages associated with its use. Many concerns stem from the crucial off-gas post-treatment stage of the process. Treatment of the off-gas can be complicated by highly abrasive feed containing abundant particulate matter. In some cases, feed with this characteristic can even damage the processor unit. The potential for air pollution due to incomplete treatment of off-gas is a persistent concern.

In many instances, the cost-effectiveness of thermal desorption can be affected by certain process limitations or site-specific conditions. Most thermal desorption activities must be performed when the soil is within a specified range of moisture content. High moisture content or high humic content can increase the reaction-time needed for process completion, resulting in a more expensive treatment operation. Site-specific materials-handling requirements can also affect cost and applications. Additionally, soil excavation and possible transportation for off-site treatment can be complicated and expensive.

Treatment of soil contaminated with metals also presents a myriad of problems. Although thermal desorption is frequently regarded as inappropriate for treating metals, it has shown some usefulness in various mercury-removal applications. Unfortunately, the leaching of mercury into water can quickly become a concern under these circumstances. Treating heavy metals can also result in the creation of a solid-residue byproduct that requires stabilization.

Other disadvantages of thermal desorption include the following:

- Clay is difficult to treat
- High concentrations of organics in the media to be treated can foul thermal desorption systems
- Excavation, crushing, screening, or storage of contaminated media can produce fugitive dust and VOC emissions
- Tars or other sticky feed can foul thermal desorption systems
- Generally higher cost

7.4.5   Design Considerations

A number of preliminary design factors are taken into consideration before the implementation of thermal-desorption treatment at a particular site. Target contaminants must be identified and on-site soil characteristics must be evaluated. The feasibility of an application will depend largely on other site characteristics, such as the vertical and horizontal extent of contamination, site layout, and current use of adjacent lands.[6]

Once identified, the contaminants targeted for removal are evaluated based on their susceptibility to treatment. Treatability tests performed at different temperatures and residence times are used to determine how effective thermal desorption will be for the removal of a particular contaminant. Several physical properties of the target contaminants are also determined during the preliminary design phase of the application.

The following physical properties of contaminants are important to consider when planning a thermal-desorption application:

- Density
- Vapor pressure
- Boiling point
- Concentration with soil

- Thermal stability
- Aqueous solubility
- Dioxin formation
- Octanol/water-partition coefficient

Moisture content of the *in-situ* soil is perhaps the most influential characteristic with regard to the feasibility of a thermal-desorption application. Most thermal desorption units can easily process soil with a moisture content less than 15%. As moisture content climbs to the 15% to 25% range, the process is slowed, but the effectiveness of the treatment is not diminished. At moisture-content levels of 20% to 25%, greater problems can be encountered that significantly impact the cost.[7] Consequently, encountering a soil with high moisture content indicates that pretreatment operations must include measures to reduce moisture in the soil. Dewatering or adding a dryer into the feed system can be used to lower the moisture content. Soils must also be protected from rain, wind, and other weather conditions that could alter the moisture content. Costs associated with extensive pretreatment operations and media-protection efforts can also significantly increase the cost of an application.

The size of the feed at a site must also be taken into consideration. Abnormally large particulate matter could adversely affect thermal-desorption operations. Particulate matter larger than 60 millimeters in diameter must be removed prior to the process to protect the equipment. The presence of excessive amounts of particulate matter at a certain site may necessitate other pretreatment measures to prevent possible complications.

In addition to feed size and moisture content, the following characteristics of soil must be considered:

- Volume of soil to treat
- Concentration of humic material
- Sieve analysis
- Plasticity
- Bulk density
- Heat capacity
- Metals concentration

Depending on the site characteristics, unique mobility or treatment capabilities may be required from a thermal desorption unit. Typically, four different types of thermal desorption units—the rotary dryer, the thermal screw, the conveyor furnace, and the asphalt plant aggregate dryer—are available. Each type of unit has its own unique process methods and components.

The rotary dryer thermal desorption unit consists of a rotating metal drum that is heated to volatilize the contaminants. The unit operates at a temperature range of 300°F to 1,200°F (150°C to 650°C). A carrier or purge gas flows concurrent with or counter-current to the soil feed. Concurrent purge gas exits the system at a high temperature, necessitating a cooling unit. Advantages of this configuration include better decontamination of fine particles and a longer residence time within the unit, since the soil is heated faster. This configuration is more effective at treating heavy petroleum products, due to the longer residence time. Counter-current purge gas exits the system at a cooler temperature, thereby eliminating the need for a cooling unit and other large downstream components associated with concurrent flow. Rotary dryer units have a wide range of treatment capacities and can be stationary or mobile.[8]

The thermal-screw thermal desorption unit uses an auger system to convey, mix, and heat soils to achieve volatilization of contaminants. This unit operates at a temperature range of 350°F to 500°F (175°C to 260°C). Lesser amounts of purge gas are needed for this unit, thus off-gas treatment operations are of a smaller scale compared to other systems. Thermal-screw desorbers are good for treating soils with high organic content, and the unit is well-suited for mobile applications. A single unit can treat from 3 to 15 tons of contaminated soil per hour.

The conveyor furnace desorber operates by conveying soil on a flexible metal belt through a heating chamber where volatilization of contaminants can occur. The unit operates at a temperature range of 300°F to 800°F (150°C to 425°C). An afterburner is typically in place to destroy organics and oxidize carbon monoxide in the off-gas. The conveyor furnace is mobile, and is capable of treating between 5 and 10 tons of contaminated soil per hour.

The asphalt plant aggregate dryer is essentially a rotary desorber with counter-current purge gas flow and no afterburner for off-gas treatment. This desorber operates at a temperature range of 300°F to 600°F (150°C to 315°C). The asphalt plant aggregate dryer can be stationary or mobile and is capable of treating between 25 and 150 tons of contaminated soil per hour.

7.4.6   Environmental Effects

The primary environmental concern when conducting thermal-desorption treatment operations is air pollution. Treatment of the off-gas generated during the desorption process can be complicated. Particulates, organic vapors, and carbon monoxide are the main target pollutants of off-gas treatment. Another potential source of air pollution is volatile metals. Thermal-desorption applications involving metals are generally inappropriate, as toxic vapors can be generated when treatment is attempted on soils contaminated with metals. Mercury emissions in particular can be very difficult to control.

Treated soil can also pose problems when it is returned to the environment. The thermal-desorption process often alters the organic composition of soil. A consequence of this alteration may be the loss of the soil's ability to support biological activity that breaks down contaminants. This is an important concern if the soil is returned to the original contaminated site. However, a covering of non-treated soil capable of supporting vegetative growth can be applied to the treated soil.

7.4.7   Results of Case Studies

7.4.7.1   Pristine Inc. Superfund Site

A full-scale cleanup using thermal-desorption technology was conducted in Reading, Ohio, at the Pristine Inc. Superfund Site. The project began on November 1, 1993, and ended on March 4, 1994. Several contaminants released to the earth by liquid-waste disposal operations from 1974 to 1981 were identified at the site. These contaminants included PAHs, VOCs, SVOCs, pesticides, and inorganic metals.

A rotary dryer was chosen as the thermal desorption unit for this project. Approximately 12,800 tons were targeted for treatment. This particular thermal-desorption application was of particular significance considering the wide range of pH and moisture content in indigenous soils. The pH was measured at levels as low as 1, while the moisture content was between 12% and 25%. Remediation objectives were set for 11 contaminants and several stack-gas emission requirements were to be met.

Thermal desorption treatment successfully met the remediation objectives for all of the target contaminants. Six of the eleven target contaminants reached levels at or below the detection limit. Additionally, all stack-gas emission requirements were satisfied. At the end of the project, the treated soil was backfilled at the site.

7.4.7.2   Anderson Development Company Superfund Site

The Anderson Development Company Superfund Site in Adrian, Michigan, was selected to undergo a full-scale cleanup via thermal-desorption technology beginning in January of 1992. Several contaminants were discovered in an unlined lagoon that had been used for waste disposal from 1970 to 1979. Among the con-taminants discovered in the lagoon were PAHs, VOCs, SVOCs, and manganese.

Approximately 5,100 tons of soil and sludge were to be treated. The soil and sludge had high moisture contents in the range of 41% to 44% after dewatering. Remediation objectives were set for VOCs and SVOCs. Of particular concern was the presence of MBOCA [4,4-methylene bis(2-chloro-aniline)] in the soil. A thermal-screw desorption unit was selected for use in this application.

The project ended in June 1993 and remediation objectives were met for all target VOCs and for MBOCA. Seven out of eight remediation goals were met for target SVOCs. The treated soil was disposed of off-site due to residual manganese contamination.

7.4.8   Engineering Economics

The fixed and variable costs for a typical thermal-desorption project include the following:

- Fixed-cost items
  - Crew and equipment mobilization
  - Treatment pad installation
  - Soil excavation/replacement

- Variable-cost items
  - Equipment leasing
  - Operation and maintenance labor
  - Utilities
  - Site supervision
  - Site quality assurance and health-and-safety support
  - Sampling analysis for process control
  - Residuals management

The cost of applying thermal-desorption technology is dependent on the scope of the treatment project. Numerous factors, including the size and location of the site, target contaminants, and local environmental regulations can profoundly affect the total cost of the project.

Fixed mobilization costs for a typical thermal-desorption application could be in the range of $10,000 to $20,000. Excavation and replacement of soil could cost around \$5 to \$10 per ton. Remediation costs are highly variable and depend on the target contaminants. Petroleum-contaminated soil could cost from \$25 to \$55 per cubic yard, and soils contaminated with other organics could cost from \$95 to \$195 per cubic yard.[9]

7.5 Incineration

7.5.1 Applicability of Technology

Incineration is a commercial treatment technology for the remediation of soils contaminated with organic wastes. Contaminants are removed by high-temperature combustion in the presence of oxygen. Soils and contaminated media are treated *ex situ*, or removed from their original location. Thermal desorption and pyrolysis are remediation technologies that are comparable to incineration, but different with regard to the goals and parameters of the combustion process. The contaminants that are susceptible to incineration are shown in Table 7.5.

Table 7.5   Contaminants Susceptible to Incineration Treatment

List of Contaminants
• Halogenated semi-volatile organic compounds (SVOCs)
• Non-halogenated SVOCs
o Heavy fuels (e.g., diesel fuel, JP5 (jet fuel))
• Chlorinated hydrocarbons
• Ordnance compounds (explosives)
• Polychlorinated biphenyls (PCBs)
• Dioxins
• Pesticides and herbicides

7.5.2 Description

The general procedure for incineration remains consistent over the broad range of contaminants and site characteristics. After preliminary design and permit acquisition are completed, the process itself typically consists of the pretreatment of media, primary incineration operations, secondary combustion, and treatment of the off-gas byproduct created during treatment. The process flowchart of an incineration application is shown in Figure 7.4.

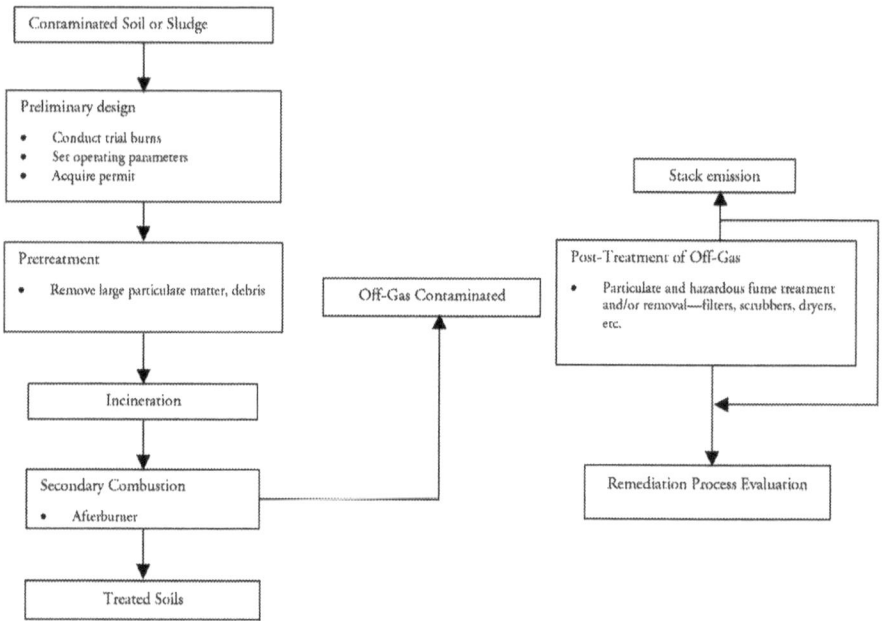

Figure 7.4   Incineration Process Flow Chart

Pretreatment of the contaminated media precedes actual incineration operations. Many incineration systems are designed to process feed within a certain size range. To avoid damage to the system, large debris and particulate matter are removed at this stage. Some characteristics of the soil, such as pH or moisture content, may be modified to accommodate treatment operations.

Primary incineration follows initial pretreatment. Contaminated media are loaded into the system and subsequently oxidized. Once oxidization is complete, incineration commences at temperatures ranging from 760°C to 1,649°C. Combustion is frequently started and sustained through the use of auxiliary fuels. Carbon dioxide and water are byproducts of this stage of the process.

Upon termination of primary incineration operations, soils and contaminated media are moved to the secondary combustion chamber. Any solid-, liquid-, or gas-phase residuals are re-heated in this stage of treatment to eliminate additional contamination. An afterburner is typically employed for secondary combustion.

The final stage of incineration treatment involves management of off-gas. Toxic gases are regularly produced during incineration treatment and must be managed.

Numerous stack-emission requirements must be satisfied to operate an incineration operation in accordance with set regulations. Several mechanisms, including baghouses and scrubbers, can be utilized to treat the off-gas.

Several types of incinerators exist, with most varieties differing in the temperatures at which they operate, and in the details of the combustion process. The circulating bed combustor operates at temperatures between 1,450°F and 1,600°F. A turbulent combustion zone is created by high-velocity air that simultaneously heats and mixes the waste. The mixing and the relatively low temperature that occurs in this particular system reduces operating costs and off-gas production.

The fluidized bed incinerator is one of the most recent models to be developed. High-velocity air is used to suspend waste as it enters the combustion chamber, effectively fluidizing it. A high-turbulence combustion zone mixes the waste and minimizes excess air within the system. A relatively long residence time provides acceptable Destruction and Removal Efficiency (DRE) and neutralizes potentially harmful residuals.[1]

A high combustion temperature characterizes liquid injector incinerators. Liquid waste is injected into the combustion chamber of the incinerator by nozzle. The waste is subsequently atomized (converted to gas), and then incinerated. Increasing turbulence or residence time can increase DRE for this system.

Infrared combustion is another method of incineration. Infrared combustors operate at temperatures of up to 1,010°C. Conveyor-belts feed waste into the combustion chamber, where electric-powered silicon carbide rods create heat for combustion. An afterburner is commonly used to destroy residuals produced by infrared combustion.

The generic commercial incinerator is the rotary kiln model, which operates at temperatures that do not exceed 982°C. A lined, rotating cylinder is used as the combustion chamber. An air-pollution control system, possibly consisting of a baghouse or scrubber, is employed to treat the off-gas produced by these incinerators.

7.5.3   Advantages and Disadvantages of Incineration

The greatest advantage of using incineration technology is the achievable Destruction and Removal Efficiency (DRE). A DRE of 99.99% is required by most regulations governing the use of incinerators, and most incinerators are capable of regularly exceeding this criterion. Additionally, DREs of greater than 99.9999% are achievable for PCBs and dioxins.

Another advantage of incineration is the speed of the treatment. Treatment by incineration is relatively rapid, with rates of treatment often ranging from 30 to 200 cubic yards of contaminated soil per day. Most incineration treatment systems are mobile. Incineration operations on-site can also reduce the amount of soil sent off for land disposal.

Incineration is often the subject of controversy, as there are many disadvantages and limitations associated with its use. It is very difficult to obtain a permit for an on-site incinerator. Extensive materials-handling requirements are often attached to incinerator use. Furthermore, transportation of hazardous wastes through populated areas to an incineration site is generally an inappropriate practice that can create a public outcry. Air pollution and toxic ash produced during incineration operations are always topics of concern as well. Residuals and off-gas byproducts must be eliminated or meticulously managed to prevent pollution in the vicinity of an incineration operation. Management of incineration operations is quite complex, as evidenced by the fact that currently only one off-site PCB incinerator is operating in the United States.

The treatment of heavy metals through incineration presents numerous problems. Toxic materials can be formed by the reaction of metals with other elements during the incineration process. Volatile heavy metals must be treated with a gas cleaning system to prevent excess air pollution. Moreover, heavy-metal treatment with incineration frequently results in the production of residuals such as incinerator ash. Sodium and potassium can also create problems for incinerators, as the ash from these elements can foul the gas ducts of an incineration system by creating troublesome particulate matter.

7.5.4   Design Considerations

A number of factors must be considered prior to the implementation of incineration for site remediation. The characteristics of site soil are of great importance. All contaminants and their respective concentrations must be determined. The soil must be classified, subjected to a sieve analysis, and undergo assessment of moisture content. Soil fusion temperature and heating value are parameters that must be known as well.

In addition to soil characteristics, a multitude of regulations governing the use of incinerators must be observed. Several laws and statutes explicitly define procedures and parameters for the construction, operation, and maintenance of incinerators. Among the laws that present guidelines for incinerator use in the United States are the Clean Air Act (CAA), the Toxic Substances Control Act (TSCA), the Noise Control Act (NCA), the Resource Conservation and Recovery Act (RCRA), and the National Pollutant Discharge Elimination System (NPDES).

Perhaps the most critical phase of the design process that involves an incineration system itself is the trial burn. The trial burn involves operating a prototype incinerator under worst-case conditions to determine its efficiency and resiliency.

Operating parameters for the system are established and tested during this stage of the design process. The system's compliance with CAA, RCRA, TSCA, and other regulations and performance requirements is monitored. Adjustments are made to ensure that the system is operating lawfully and achieving the DREs required. During the trial burn, the incinerator must prove that it is capable of achieving 99.9999% DRE for PCBs and 99.99% for other constituents. Risk assessments and emissions monitoring are also performed throughout the design process. Upon completion of this extensive series of testing and monitoring, a permit application for the incineration operation is submitted. If the design is approved and a permit is granted, the incineration operation can begin.

7.5.5   Environmental Effects

The potential of creating air pollution through stack gas emissions is the most significant threat posed to the environment by incineration. Off-gas stack emissions must be closely monitored for the duration of any incineration operation. Many emissions can contain particulates along with toxic or carcinogenic gases or both. Furans, PAHs, herbicides, and partially-burned polyvinyl chloride are gases that could create harmful air pollution. To prevent the release of contaminated "off-gas," several different mechanisms can be employed as part of the incineration system. Off-gas is often treated through the use of a baghouse, scrubber, wet electrostatic precipitator, or spray dryer.

Another concern is the production of hazardous residuals. Some solids will not be fully destroyed during the incineration process and will remain as tangible hazardous materials. Secondary combustion chambers are installed to prevent excessive residual production. Afterburners are typically used to reheat contaminated media that has not been destroyed in the primary combustion chamber. Residuals that cannot be eliminated or treated are generally subject to land disposal.

7.5.6   Results of Case Studies

7.5.6.1   Coal Creek Superfund Site

From 1949 to 1983, maintenance and scrapping of electrical equipment contaminated several waste-disposal areas at Coal Creek in Chehalis, Washington. Incineration was selected for use in remedial action at the site that began in January of 1994.

Approximately 9,175 tons of soil was targeted for remediation. A rotary kiln incinerator with a secondary combustion chamber was selected for use at the site.

The soil was contaminated with PCB, lead, and other metals including copper, mercury, and zinc. The standard DREs of 99.9999% for PCB and 99.99% for other constituents were used as remediation objectives.

The incineration operations lasted from January 1994 to May 1994. Data indicated that all DRE requirements and relevant emissions criteria were met. The total cost of incineration at the Coal Creek Superfund Site was $8.1 million.

7.5.6.2   Former Nebraska Ordnance Works

Incineration was used for remedial action at the Former Nebraska Ordnance Works in Mead, Nebraska. Rinse-water discharges and the burning of explosives on site from 1942 to 1959 resulted in the contamination of local sediments.

A rotary kiln incinerator with a secondary combustion chamber was selected to remediate approximately 16,440 tons of contaminated soil and debris. Explosives and propellants were the contaminants, including TNT contamination of up to 133,000 mg/kg, RDX contamination of up to 23,270 mg/kg, TNB contamination of up to 430 mg/kg, and DNT contamination of up to 119.3 mg/kg.

The standard DRE of 99.99% was used as the remediation objective. Specifically, TNT was to be reduced to 17 mg/kg, RDX to 5.8 mg/kg, TNB to 1.7 mg/kg, and DNT to 0.9 mg/kg.

The Former Nebraska Ordnance Works project was completed in a very short time, lasting from September 1997 to December 1997. Process data along with post-project soil sampling and analyses indicated that all remediation objectives were met. Following the completion of remedial action, the incinerator was demobilized and removed from the site by May 1998. The total cost of incineration at the Former Nebraska Ordnance works was approximately $10.7 million. This cost equates to roughly $394 per ton of contaminated media treated.

7.5.7   Engineering Economics

Cost considerations for the incineration method of dealing with waste tend to be the reason why it is not a widely applied methodology. Some of the major fixed and variable expenses for this method are shown below.

- Fixed-Cost Items
  - Crew and equipment mobilization
  - Treatment-pad installation
  - Soil excavation

- Variable-Cost Items
  - Equipment leasing
  - Operating and maintenance labor
  - Utilities
  - Site supervision
  - Site quality control and health-and-safety support
  - Sampling and analysis for process control

The bulk of incineration expenses are related to the mobilization of the incinerator. Regardless of the amount of soil to be treated and the extent of contamination, mobilization cost is a fixed-cost item. Mobilization can require an expenditure of between $300,000 and $500,000. Treatment costs per cubic yard are often in the range of $100 to $500.[2] An all-inclusive breakdown of costs approximates off-site incineration costs at $200 to $1,000 per ton of contaminated media treated. Expenses are significantly increased if PCB-contaminated media are to be treated. Off-site incineration of PCB-contaminated media can cost between $1,500 and $6,000 per ton.[3]

7.6 Bioremediation

7.6.1 Applicability of Technology

Bioremediation is the use of living organisms for the remediation of contaminated soil, sediment, or water. Several technologies have been successfully applied in this field, and many others have been proposed. Often, the potential for a technology is unappreciated until an appropriate site can be identified. For site bioremediation, a technology might address only one component of a complex mixture, or might attempt a comprehensive cleanup. A site may be as uncomplicated as a closed gasoline station, or as complex as a contaminated aquifer.

Table 7.6   Accelerated Bioremediation Applications

- Groundwater Plumes
    - o   Permeable bio-barriers (migration control)
    - o   Plume treatment (depends on size/coat)
- Source Areas
    - o   Migration control or source treatment
    - o   Bioaugmentation
- Vadose Zone
- Bioventing or Cametabolic Venting

Table 7.7   Chemical Classes And Their Susceptibility To Bioremediation

Chemical Class	Susceptibility to Bioremediation
Petroleum hydrocarbons	Readily biodegradable
Industrial solvents	Somewhat difficult to biodegrade
Wood preservatives	Readily biodegradable
Pesticides	Difficult to biodegrade
PCBs	Some evidence; not readily biodegradable
Metals	Not biodegradable
Radioactive materials	Not biodegradable
Corrosives	Not biodegradable
Asbestos	Not biodegradable

Table 7.8   Biodegradability Levels of Chemical Classes

Level	Levels of Biodegradability
#1	**Very Easy**—Petroleum-related (natural compounds). Examples include crude oil, gasoline, and diesel fuel.
#2	**Easy**—Solvents, wood preservatives, polynuclear aromatic hydrocarbons, petroleum wastes, chemical manufacturing wastes/products, many pesticides and paint solvents. Trichloroethylene (TCE), perchloroethylene (PCE), trichloroethane (TCA), polychlorinated biphenyls (PCBs), complex polynuclear aromatic hydrocarbons (>5 rings), trinitrotoluene (TNT), DDT, and dioxin.
#3	**Difficult**—Metals (may alter oxidation and reduction, and may be toxic), salts, highly insoluble compounds, synthetic oils, and polymerized asphaltenes.

7.6.2   Description

There are two basic types of bioremediation. Biostimulation provides nutrients and suitable physiological conditions for the growth of indigenous microbial populations. This promotes increased metabolic activity, which then degrades contaminants. Bioaugmentation involves the introduction of specific blends of laboratory-cultivated microorganisms into a contaminated environment or into a bioreactor to initiate the bioremediation process.

Microbial biodegradation is an innovative, emerging technology for handling subsurface water contamination. It is a natural process that can be accelerated by the injection of certain nutrients such as dissolved oxygen, nitrates, and acetate. Studies by the U.S. Environmental Protection Agency have shown that this strategy can result in the complete removal of contaminants, whereas other proposed restoration strategies are not as effective. Biodegradation technologies are currently being studied at several U.S. Department of Energy laboratories in an effort to remove or contain volatile organic compounds.

Microorganisms, and to a lesser extent plants, have the capacity to degrade or transform many types of contaminants. Bioremediation has been defined by the American Academy of Microbiology as the use of living organisms to reduce or eliminate environmental hazards resulting from accumulations of toxic chemicals or other hazardous wastes. Bioremediation can be natural or accelerated. Natural bioremediation relies on naturally occurring microbial and plant processes. Accelerated bioremediation occurs via the addition of amendments such as nutrients and electron acceptors, or microorganisms, including genetically engineered microorganisms (GEMS), or by manipulating physical, chemical, or hydrological processes.

Bioremediation may be achieved either in situ at the surface or within the sub-surface—or *ex situ*—above ground.

7.6.2.1   In-Situ/Surface and Subsurface Bioremediation

In-situ bioremediation occurs at the original site of contamination and without excavation or removal of soil. Examples of *in-situ* bioremediation include bioventing, air-sparge systems, sponge-vent systems, and sparge barriers. If the contamination is in the upper 12 inches of the soil, treatment can consist of tilling to provide aeration, and adding nutrients and water to promote bacterial growth. Since oxygen is not very soluble and is easily depleted, it may be necessary to deliver oxygen to the contaminated area for the respiration process. This is done by withdrawing groundwater, adding an oxygen source such as air, pure oxygen, hydrogen peroxide, or ozone, and re-injecting the water using injection wells or trenches.

7.6.2.2   Ex-Situ/Aboveground Bioremediation

Ex-situ bioremediation occurs when contaminated material is removed and transferred to a separate treatment site or facility. Examples of *ex-situ* bioremediation include composting, biopile treatment, bioreactors, and landfarming. Slurry-phase treatment is a process that combines contaminated soil with water to create a slurry, which is broken down in a bioreactor. If necessary, nutrients and oxygen can be added. The soil and water are separated after treatment. In solid-phase treatment, contaminated soil is placed in a treatment bed where nutrients, moisture, and oxygen are added to promote decomposition.

The majority of contaminants that have been treated by bioremediation to date are petroleum derivatives, including fuels, petroleum solvents such as acetone and ketone, and polycyclic aromatic hydrocarbons (PAHs), which are found in coal tars and creosotes. In general, the most easily treatable group tends to be short-chain hydrocarbons and single-ring aromatics, with difficulty increasing as one moves towards the longer-chain hydrocarbons, chlorinated aromatics and PAHs. Degradation pathways include:

- Aerobic pathways: oxidation where oxygen is the terminal electron acceptor and producer of specific enzymes
- Anaerobic pathways

Table 7.9   Aerobic and Anaerobic Degradation of Chemical Classes

Aerobic Degradation	Anaerobic Degradation
• BTEX (aromatic hydrocarbons: benzene, toluene, ethylbenzene, xylene) • Ketones, esters (acetone) • Petroleum hydrocarbons • Polyaromatic hydrocarbons (creosote) • Chlorinated solvents (TCE, PCE) • Organic cyanides	• BTEX • Ketones, esters • Chlorinated solvents (TCE, PCE)

7.6.3   Advantages of Bioremediation

There are many advantages to bioremediation when compared to other technologies. For the past decade, pump-and-treat systems were the preferred method for soil and groundwater cleanup. These systems consist of a series of wells used to pump water to the surface, and a surface treatment facility used to clean the extracted water. Recent studies have shown that since many common contaminants tend to become trapped in the subsurface, the pumping of extremely large volumes of water over very long periods of time may be required in order to completely flush them out of the soil. *In-situ* bioremediation, which treats contaminants in place instead of requiring their extraction, may speed the cleanup process. Consequently, bioremediation is likely to take a few months or years to reduce the contaminants' concentration in soils.

Potential advantages of bioremediation, when compared to other *in-situ* methods, include destruction rather than transfer of the contaminant to another medium, minimal exposure of on-site workers to the contaminant, long-term protection of public health, and possible reduction of treatment costs.

Bioremediation can be used to clean up sites with leaking underground storage tanks (USTs) that have contaminated both soil and groundwater with aromatic hydrocarbons, when other conventional technologies are not successful. For example, when a UST is located beneath a building, which prohibits excavation of contaminated material, *in-situ* bioremediation can be used without disturbing the structure. This technology is also successful for remediation of releases that have affected a railroad bed and track. Conventional remediation would involve excavating the contaminated soils, causing delays and loss of revenue. *In-situ* bioremediation would allow the treatment of the impacted soils without any interruption of rail service, while realizing a significant overall cost savings.

Other than keeping site disruption at a minimum, there are other advantages of bioremediation over conventional excavation and landfilling techniques. Microbial processes permanently destroy contaminants, eliminating long-term liability exposure that may persist when using non-destructive treatment methods. *In-situ* remediation also eliminates the liabilities and costs associated with transportation of excavated soils. In general, biological systems are often less expensive to implement, as they have no moving parts and require no artificial energy source, thereby reducing overall remediation costs. Moreover, bioremediation can be coupled with other treatment techniques into a treatment train.

As with any treatment technology, bioremediation also has some limitations. Although bioremediation is effective for petroleum compounds, it is not applicable to highly chlorinated compounds and metals. Another drawback is that the breakdown products of bioremediation may be more toxic than the original target compound. For example, trichloroethylene (TCE) can be broken down into vinyl chloride, a known carcinogen. Further, there is a need for long-term, extensive monitoring of the bioremediation project that would not be required for excavation-and-disposal techniques. Monitoring would include visiting the site on a routine schedule to assess the status of the biodegradation and ensure that all necessary constituents required by the microbes are available.

Despite these limitations, the many advantages of bioremediating petroleum-contaminated soils and groundwater will cause the continued growth of this technology as a remedial option. Bioremediation offers the ability to remediate the contaminant in place, has the potential to completely break down contaminants into innocuous end-products, and is very cost-effective. As more bioremediation projects are completed successfully, the technology will undoubtedly gain acceptance as a routine treatment method. In situations where bioremediation has gained regulatory approval, this technology has been shown to reduce costs, avoid use-interruption of affected sites, and provide an environmentally sound method of site cleanup.

Table 7.10 Advantages of Bioremediation

Environmentally safe, natural process Usually does not produce toxic byproducts Offers long-term solution for balanced ecosystem
Does not disrupt normal site operations Can be used where the problem is located on-site
Cleanup occurs in place Contaminants are not transferred or converted to another phase
Sophisticated yet easy-to-use technology Application is simple
Low-cost and efficient Consumes less energy Can accelerate dissolution and reduce cleanup time

Table 7.11 Disadvantages of Bioremediation

Potential for system fouling Operation and maintenance costs
Toxic by-products be formed in the treatment process Negative geochemical changes Highly specific reactivity for contaminants Potential to form undesirable degradation intermediates
More time is often required relative to traditional methods Other actions, such as incineration, are faster Not all biodegradable compounds can be rapidly and completely degraded

Requires careful monitoring to ensure effectiveness
May be unfeasible for large dilute or concentrated plumes
Typically required nutrient delivery and mixing
More research is needed
For locations with complex mixtures of contaminants
To develop appropriate technologies, establish predictability, reliability and effectiveness parameters for specific compounds under various conditions.

7.6.4   Design Consideration

- Basic Steps
    - Site investigation
    - Free product recovery
    - Microbial degradation study
    - System design
    - Operations
    - Maintenance

- Site Parameters
    - Direction of groundwater flow
    - Rate of groundwater flow
    - Depth of water table
    - Depth of contaminated soil
    - Specific yield of aquifer
    - Heterogeneity of soil
    - Hydraulic connection within aquifers
    - Potential recharge and discharge zone
    - Seasonal fluctuation of water table

- Investigation and Feasibility Studies
  - ○ Monitoring wells
    - Water level
    - Conductivity
    - Presence of hydrocarbon
    - Mapping of plume

  - ○ Hydrogeological investigation
    - Feasibility test

- Hydraulic Requirement
  - ○ Injection and circulation of nutrients (if necessary)
  - ○ Continuous nutrient and oxygen injection (if necessary)

- Choosing bioremediation strategy

Engineering Parameters for Site Remediation

- Chemical characteristics
- Hydraulic properties of aquifers
- Site characterization factors

Among the most prevalent problems today in the environmental arena is the contamination of soils and groundwater with petroleum compounds. Because of the large number of petroleum-contaminated sites requiring cleanup and the cost involved with the conventional approach to excavation and landfilling, a need exists to develop new remedial technologies for such sites. Bioremediation is a technology that has been increasingly refined and developed.

Many factors influence the proper choice of an engineered bioremediation system. These include the type and extent of contamination, site characteristics, cleanup goals, and economics. Therefore, it is not possible to apply a single remedial approach to every situation. Prior to implementing a bioremediation project, a multidisciplinary approach to characterizing hydrogeologic, geologic, chemical, and microbiological characteristics should be undertaken. Key microbiological considerations include carbon and energy sources, electron acceptor availability, temperature, nutrients available to the microbes, soil or groundwater pH, and soil moisture content.

Petroleum and petroleum products such as gasoline, fuel oil and diesel fuel are complex mixtures of organic compounds. Gasoline, for example, contains over hundred different substances, which can hinder biodegradation since no single microbe can degrade all of the substances. Therefore, during site characterization, an evaluation of the indigenous microorganisms available to perform biodegradation of the petroleum may be required. From this information, the overall effectiveness of the remediation can be evaluated to determine whether additional microbes are necessary to achieve cleanup goals.

Once a petroleum compound enters into the soil or groundwater, indigenous microbes perform natural biodegradation. However, several limiting factors slow the rate of biodegradation, allowing the petroleum to persist and continue to migrate, thus causing further contamination. Two of the most important limiting factors are the availability of oxygen and the availability of nutrients for use by the microbes. Without ample supplies of oxygen and nutrients, the amount of contamination removal that can occur is limited.

The most common bioremediation method utilizes indigenous microorganisms to degrade the petroleum. Other methods include the introduction from another area of microorganisms capable of degrading the particular contamination and, in some cases, genetically engineered microbes designed to degrade the compound of interest. When indigenous microbes are selected, and after all preliminary investigation work is complete, it is generally possible to stimulate the microbes to begin using the petroleum compounds as a food source at a much greater rate by compensating for limiting parameters, such as by introducing additional oxygen or nutrients.

Even simple steps such as the addition of fertilizer or oxygen will provide the required nutrients for the microbes. This will allow the microbes to multiply at a rapid rate while consuming the petroleum compounds. It has been shown that the addition of fertilizer increases the metabolic rates of the petroleum-degrading microorganisms, which can increase the removal rates of petroleum compounds as much as tenfold.

The time constraints involved in achieving cleanup goals are another limitation which greatly influences whether bioremediation can be the technology of choice. Depending on the circumstances of the remediation project, it could take years to complete. In addition, there are regulatory restraints. Bioremediation has not been accepted by all state regulatory agencies, as it is perceived as an unproven technology. As a result, some agencies are reluctant to approve the technique.

In deciding on the best system for site remediation, careful appraisal of the following aspects is important:

- Water is a basic requirement for bacteria. Water provides bacteria with a means of transport as well as with oxygen. Biotreatment is optimum at 20% to 80% soil saturation, although it remains efficient even at 100% saturation.

- Soil type is a major consideration for bioremediation, particularly for the *in-situ* process. Tight soils such as clays and silts limit the movement of water as well as the amount of oxygen, therefore making it difficult to treat. Gravel and sandy soils are ideal for penetration of water, oxygen, bacteria, and nutrients. Soil type will also dictate *ex-situ* methodology.

- Optimum temperatures for bioremediation are between 20°C and 40°C. Temperatures below 20°C slow the process by making it difficult for the bacteria to reproduce, while temperatures above 40°C can destroy bacteria, or cause them to mutate.

- A pH Range of 5 to 9 is ideal. However, if the pH is higher or lower, it can be adjusted by adding a buffer material. The addition of a buffer material balances the pH and is beneficial to the microorganisms and the health of the soil.

- Contaminant type and concentration are important considerations that must be factored in to determine the specific bacteria and necessary nutrients required.

- Oxygen levels must be high enough for the bacteria and for the breakdown of hydrocarbons. Oxygen can be supplied through mixing, water, pressure wells, mechanically, or through chemical means where required. Bacteria will become anaerobic when oxygen levels are low. Anaerobic bacteria consume material at a rate approximately ten times slower than aerobic bacteria.

7.6.5 Environmental Effects

The environment in which the contaminated site exists influences the type of organisms that can be used. For example, in cold environments (0° to 15°C) psychrophilic (cold-loving) organisms would be effective; conversely, in hot environments (>45°C) thermophilic (warmth-loving) organisms would be effective. Temperature, pH, heavy-metal concentration, microflora, and microbial diversity are some of the environmental factors that must be considered. If practical, the environment may be adjusted to provide optimum conditions.

Assuming that the scientific obstacles are overcome, the safety of releasing genetically modified organisms into the environment is currently under discussion. To avoid any unforeseen problems, one strategy under investigation is to make the organism dependent on the contaminant for life; therefore, when the contaminant has been successfully removed from the environment, the organism will die. Another option is to take contaminated soil and water to contained facilities for biological decontamination, which would be a much more expensive proposition.

The long-term effects of introducing non-native bioremediative organisms into an ecosystem, such as in bioaugmentation, are not currently well understood, and some of these effects could ultimately prove more harmful than the original pollutant. The environmental damage done by an introduced bacterium could be more harmful than the damage done by pollution.

7.6.6   Results of Case Studies

7.6.6.1   Remediation of BTEX and TPH

During construction at a service station, it was discovered that gasoline and diesel fuel were present in the soil and groundwater. The underground storage tanks and impacted soils were later removed. A temporary groundwater treatment system was installed to treat localized contaminated groundwater. Upon further assessment, an estimated 700 cubic yards of soil and associated groundwater were found to be contaminated with BTEX and TPH concentrations, exceeding Pennsylvania Department of Environmental Resources (PADER) standards.

A groundwater-monitoring program was implemented to monitor BTEX and TPH levels, and an on-site bioremediation program was initiated. An existing monitoring well in the contaminated area was inoculated with M-1000HR and OSNF nutrients.

Quarterly sampling showed a reduction in BTEX of over 85%. After the last sampling round, the owner petitioned PADER to close the site. Three years later, PADER determined that no further action on the site was required.

7.6.6.2   Bioremediation—#2 Fuel Oil *In Situ*

A Massachusetts residential property was found to contain #2 fuel oil underneath the basement floor. The oil release came from a leaking aboveground storage tank located next to the foundation. A subsequent site investigation found TPH (total petroleum hydrocarbon) levels as high as 4,500 ppm in the soil.

Cost estimates to remove the foundation and basement floor, as well as the contaminated soil, ranged from $150,000 to $200,000. The owner selected bioremediation as an alternative solution for elimination of the fuel-oil contamination. A series of six-inch-diameter holes were drilled through the basement floor, and M-1000HR, OSNF #1, nutrients and water were metered into the holes to assure thorough penetration and proper soil saturation. Moisture and nutrient levels were monitored and maintained. After five months, TPH levels were reduced to less than 500 ppm, except in one small area where hardpan soil lenses prevented proper saturation. The concrete floor was removed in this area, the soil broken up, and the area treated with bacteria and nutrients.

Contaminant levels in the problem area immediately started to drop following treatment, and by the end of seven months, that area also showed TPH values well below the target 500 ppm. The next month, confirmation testing showed that all areas complied with state clean-up standards. The total cost to remediate the site was less than $25,000.

7.7 Adsorption

7.7.1 Application

Adsorption is widely employed in a number of pollution-abatement programs as an integral part of water- and waste-treatment processes. When used in water and waste treatment, the overall main objective is to remove toxic organic compounds, thus rendering the purified water suitable for discharge or reuse. Adsorption is also used in the purification of gases as an air-pollution control strategy. In this discussion, however, adsorption will be discussed as it relates to the removal of objectionable or toxic organic compounds from waste.

Various materials such as activated carbon and synthetic resins are suitable adsorbent materials for the adsorption process. Adsorption of toxic compounds onto activated carbon has been demonstrated to be feasible and reliable for the removal of organics from water and waste.[1, 2] Adsorption onto synthetic resin has proved equally suitable for water and waste treatment.[3-7] Synthetic resin has been demonstrated to be superior to activated carbon for industrial waste treatment,[8, 9] or competitive at high concentrations.[10] The most remarkable feature of synthetic resin is its ability to desorb completely, as opposed to carbon. This quality of resins to desorb completely is an important engineering consideration for industrial applications where recovery may be of interest.

However, activated carbon is more commonly used because it has been demonstrated to be adequate for the removal of a variety of organic contaminants found in water or waste. Moreover, it is less expensive than synthetic materials. Although this section centers on the use of activated carbon, it must be stressed that most of the discussion is applicable to the use of other adsorbent materials as well.

7.7.2   Theory

The adsorption phenomenon involves the (accumulation) of substances at an interphase (surface). The process may occur between two phases such as liquid-liquid, liquid-solid, gas-solid, and gas-solid interphases. A liquid-solid-phase system is usually the case for water and waste treatment systems. The adsorbing phase is generally termed the adsorbent, while the substance being adsorbed is termed the adsorbate.

The primary driving force for adsorption may be a lyophobic (solvent-disliking) character of the solute relative to the particular solvent, or of a high affinity of the solute for the solid.[11] In this regard, it can be stated that the more hydrophilic the molecule, the less likely it will move to the interphase for adsorption to occur. It is believed that the combination of the above two forces contribute to the adsorption phenomenon.

There are basically three types of adsorption interaction: adsorption due to electrical attraction, adsorption due to Van der Waals forces, and chemical adsorption. Adsorption due electrical attraction is predominant in the ion exchange process, and involves the concentration of ions at sites due to electrostatic attraction. Adsorption due to Van der Waals forces is generally termed physical adsorption, and in this case, once the molecule is adsorbed to a site, it can still migrate or undergo translational movement within the adsorbent. Chemical adsorption is the term used when the adsorbate interacts chemically with the adsorbent, and in this case adsorbed molecules are not free to move to another site within the interphase. In general, all three types of adsorption may occur simultaneously, and in most adsorption processes for the removal of organics from water or waste, physical adsorption is believed to play a dominant role.

7.7.3   Factors Affecting Adsorption

The factors that affect the adsorption process can be broadly grouped into three, namely:

- adsorbate characteristics
- adsorbent characteristics
- solution characteristics

7.7.3.1 Adsorbate Characteristics

The adsorbate properties that have been recognized to affect the adsorption process include functionality, branching or geometry, polarity, hydrophobicity, dipole moment, molecular weight and size, and aqueous solubility.[12] A study of a series of sulfonated alkylbenzene indicates that optimum adsorption capacity increases with molecular weight.[13] In a related study[14] on the adsorption of 93 petrochemicals, it was observed that as molecular weight and polarity increases, and as solubility and branching decrease, adsorption capacity increases. In addition, functionality was observed to have a substantial effect on the amount adsorbed. Other studies have shown similar findings on the effect of molecular weight and functionality on the adsorption of series of compounds on activated carbon.[15]

7.7.3.2 Adsorbent Characteristics

The properties of an adsorbent generally affect the capacity and rate of adsorption. In this regard, surface area, pore size, and particle size are most important. Adsorption is a surface phenomenon; hence, the larger the surface area, the higher the adsorption capacity. Pore size plays an important role because molecules with diameters larger than the pore diameter are excluded from the pore system by size alone.[6] More importantly, if the adsorbent is composed of macropores, transition pores, and micropores, the rate of adsorption may be different for each pore region. In general, the lower the particle size, the higher the rate of adsorption.[16] This rate has been determined to be inversely proportional to the square of the adsorbent particle diameter.

7.7.3.3 Solution Characteristics

Aqueous characteristics such as temperature, pH and the presence of other competing substances may influence the adsorption process. Adsorption reactions are generally termed exothermic; as such, increasing the aqueous temperature reduces the optimum capacity, although the rate of adsorption is increased. In general, adsorption of typical organic pollutants from water and waste is increased with decreasing pH.[11] The presence of other substances in a waste sample may affect the adsorption of a particular compound, and in general, the presence of other organics in solution usually reduces the adsorption capacity for a particular organic compound.[6]

7.7.3.4   Design Consideration

The use of adsorption systems for waste treatment involves various activities, including waste characterization, feasibility studies, and engineering design of the system. As shown in Figure 7.5, the process involves bringing the waste containing the toxic organic compounds into contact with the adsorbent. As the stream passes through the adsorbent, the toxic organic compounds are adsorbed onto the adsorbent.

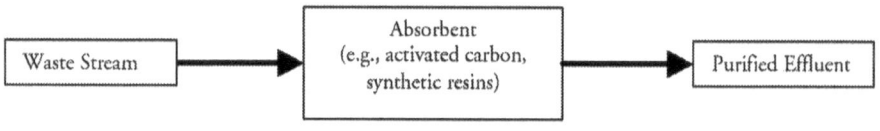

Figure 7.5   Adsorption Process

Therefore, the efficiency of such a system must depend on the characteristics of both the waste stream and the adsorbent. Understanding of all the interactions involved can only be achieved through treatability studies or bench-scale laboratory studies.

Waste Characterization

The benefits of detailed and accurate waste characterization are several. Knowledge of its chemical constituents will aid in the design of the feasibility study and other treatment unit processes that may be required. It may also indicate whether the adsorption system should be used as part of end-of-pipe treatment, or only for isolated streams. Important parameters to determine are:

• Identity and concentration of all organic constituents
• Identity and concentration of all inorganic constituents
• pH
• Suspended solids
• Dissolved solids
• Oil and grease

Identity and concentration of all organic constituents indicate the quantity of organics present in the waste. Total organic carbon measurement may also be used to obtain the quantity of organics present. Suspended solids, pH, and oil-and-grease concentrations will aid in determining whether pretreatment is required. For example, a water stream with a high concentration of suspended solids will require filtration as a pretreatment process if clogging is to be avoided. Various organic compounds have an optimum pH range for adsorption; hence, pH adjustment may be required prior to the adsorption process. The inorganics present, with their respective concentration levels, may indicate possible air-pollution problems during thermal regeneration of the activated carbon. Once the above information is known, the next step is to determine the feasibility of using the adsorption process.

7.7.4 Feasibility Study

Having discerned the waste characteristics, the next step is to determine the type of adsorbent required. In the case of a waste stream with one major organic compound, and where recovery is desired, a synthetic resin such as XAD4 or XAD2 may be considered. Although synthetic resin is more expensive than activated carbon, reversible adsorption occurs at a higher rate with synthetic resin than with carbon. However, the levels at which the organic compounds are present in the waste indicate that recovery is not attractive. Therefore, carbon can be used, since it is cheaper than resins and has been found suitable for the adsorption of a variety of organic compounds.

Because several types of activated carbon are manufactured, it may not be obvious at first which one to use. However, information can be gathered from similar industries with similar waste characteristics, from activated-carbon manufacturers, or from environmental-engineering consultants. Based on this information, at least three types of carbon should be selected for laboratory screening.

The basic studies required prior to the design of fixed-bed adsorber are equilibrium studies, column studies, and regeneration/reactivation evaluation.

Equilibrium Studies

Equilibrium studies indicate the feasibility of using activated carbon to adsorb toxic organic compounds, and the ultimate capacity for such compounds. Batch tests (bottle techniques) are usually employed, and involve bringing known volumes of waste—whose chemical constituents and concentrations are known—into contact with known quantities of carbon. This type of experiment usually requires the use of series of batch reactors. The aim is to produce equilibrium concentrations of the contaminants both in the aqueous phase and on the carbon.

The experiment is usually designed to cover the concentration range expected in the raw waste.

The equilibrium relationship between the contaminant concentrations in each of the two phases (liquid and solid) as mentioned above is called the equilibrium isotherm. Adsorption isotherms can be expressed mathematically, but the functional relationship depends on the system. The Freundlich adsorption equation, as shown in equation 7.1 adequately describes most adsorption isotherm data, especially for single-component systems.

$$X/M = KC^{(1/n)} \quad \text{(Equation 7.1)}$$

where
X = amount of organic compound adsorbed
M = mass of carbon
C = equilibrium concentration of organic compound
K = equilibrium constant
n = equilibrium constant

The above equation may also be suitable for the description of a multi-component system at low concentrations, and when a group parameter such as total organic carbon is used as the measure of concentration level. However, for multi-component systems, multi-component adsorption systems should be utilized.[16, 17]

In practice, equation 7.1 is most useful when plotted on logarithmic paper. Therefore, taking the logarithm of equation 7.1 yields:

$$\text{Log } X/M = \log K + 1/n \log C \quad \text{(Equation 7.2)}$$

It is readily seen that this resulting equation is of the linear type, with an intercept of K and slope 1/n. In general, K is indicative of the capacity, while 1/n is indicative of the adsorption intensity. Table 7.12 presents these two parameters for the toxic organic compounds of interest.[18] These data resulted from studies using pulverized Calgon Filtrasorb 300 activated carbon, at initial concentration of 1.0 mg/l, and at the indicated pH. These parameters can be used to determine the capacity of the organic compound on Filtrasorb 300 activated carbon using equation 7.1 at a desired-equilibrium aqueous concentration. On the other hand, these parameters are indicative of various interactions involved, which are the consequences of adsorbent, adsorbate, and aqueous characteristics. For example, at pH 5.3, the value of K for carbon tetrachloride is reported to be 11.1 mg/g.

However, the value is reported (Table 7.12) to be approximately 40.0 mg/g at pH 7.0, even though l/n remains practically constant. This indicates that the feasibility study must be well designed, so as to determine the optimum conditions (the liquid pH in this case) for adsorption. The isotherm parameters are also very sensitive to temperature changes; hence, isotherms should be conducted at the temperature expected during the actual application.

Table 7.12   Freundlich Parameters.[18]

Toxic	Pollutant	No and Name	pH	K (mg/g)	l/n
6	Carbon tetrachloride		5.3	11.1	0.83
			7	40	0.84
8	1,2,4—trichlorobenzene		5.3	157	0.31
10	1,2—Dichloroethane		5.3	3.6	0.83
11	1,1,1—Trichloroethane		5.3	2.5	0.34
14	1,1,2—Trichloroethane		5.3	5.8	0.6
21	2,4,6—Trichlorophenol		3	219	0.29
			6	155	0.4
			9	130	0.39
23	Chloroform		5.3	2.6	0.73
			7	11	0.84
24	2—Chlorophenol		3.0–9.0	51	0.41
25	1,2—Dichlorobenzene		5.5	129	0.43
26	1,3—Dichlorobenzene		5.1	118	0.45
27	1,4—Dichlorobenzene		5.1	121	0.47
29	1,1—Dichloroethylene		5.3	4.9	0.54
31	2,4—Dichlorophenol		3	147	0.35
			5.3	157	0.15
			9	141	0.29

Toxic	Pollutant	No and Name	pH	K (mg/g)	l/n
37	1,2—Diphenylhydrazine		5.3	16,000	2
38	Ethylbenzene		7.3	53	0.79
44	Methylene chloride		5.8	1.3	1.16
48	Dichlorobromomethane		5.3	7.9	0.61
			7	19	0.76
54	Isophorone		5.5	32	0.39
55	Naphthalene		5.6	132	0.42
57	2—Nitrophenol		9	85	0.39
58	4—Nitrophenol		3	80	0.17
			5.4	76	0.25
			9	71	0.28
64	Pentachlorophenol		3	260	0.39
			7	150	0.42
			9	100	0.41
65	Phenol		3.0–9.0	21	0.54
66	Bis(2-ethylhexyl) phthalate		5.3	11,300	1.5
67	Butyl benzyl phthalate		5.3	1,520	1.26
68	Di-n-butyl phthalate		3	220	0.45
78	Anthracene		5.3	376	0.7
85	Tetrachloroethylene		5.3	51	0.56
86	Toluene		5.6	26	0.44
87	Trichloroethylene		5.3	28	0.62
			7	21	0.5

As indicated earlier, batch studies are generally used to develop equilibrium data. These data can then be applied to the design of fixed-bed absorbers.

However, it has been demonstrated that the capacities developed with batch procedures may be different from observed column capacities.[10, 18, 19] Although there
is no theoretical basis for this difference, it is still important to consider. A survey
of industrial applications of aqueous-phase activated carbon adsorption for the
control of pollutants from the manufacture of organic compounds clearly show
(Table 7.13) that in many applications, capacity obtained from batch data may
differ from that obtained in the fixed-bed adsorbers.[20] It has been pointed out
that non-ideal flow conditions and limited contact time employed in column
operation may play an important role in reducing column capacities to values
lower than those predicted from the batch data. In all cases where the column
capacity was found to be different from batch results, it was observed that the
column capacity was lower than the batch results. As such, a compound's adsorb-
ability as indicated by the batch results may not necessarily predict removal in a
dynamic mode of operation. Therefore, it is suggested that column procedures be
used to develop the equilibrium capacity that is required in the design of fixed-
bed adsorption systems.

**Table 7.13   Comparison of Activated Carbon capacity in Isotherms
test vs. capacity in column test.[20]**

Waste Composition	Capacity (g/g)	
	Isotherm	Column
Petrochemical plant biotreated efflent (as COD)	0.650	0.24
Petrochemical plant biotreated effluent (as COD)	1.300	0.26
Petrochemical plant raw waste(as COD)	0.075	0.05
1,2—Dichloroethane	0.208	0.17
1,2—Dichloroethane	0.910	0.44
Methyl ethyl ketone	0.160	0.13
Butyraldehyde	0.220	0.19
4—Component mixture (as TOC)	0.170	0.15
Refinery effluent (as COD)	1.750	0.30

Column Studies

The primary purpose of column studies is to determine the parameters needed for the design of the adsorption system. Fixed beds are generally employed; hence, the role of a column in these studies is to mimic the actual operational mode. Basic parameters to be established are:

- Contact time
- Carbon usage rate
- Pretreatment requirement

Contact time defines the amount of carbon required in the adsorption column, and can be used for the estimation of capital cost for the system. Carbon usage rate indicates how often the carbon will be replaced, and reflects on the operational cost for the system. Pretreatment requirements are generally defined after waste characterization.

A column experiment can be conducted by passing waste with known concentration, C_0, through a stationary adsorber. Initially, the waste comes in contact with fresh carbon, which adsorbs almost all of the contaminants. Contaminants that are not adsorbed in the first layer are subsequently adsorbed as they travel through the adsorption column. At this time, the adsorption zone is located near the influent end of the column. However, once this layer has become fully saturated with the contaminant, it moves through the column to regions less saturated with the contaminant. This zone is generally termed the wave front or mass transfer zone. As the wave moves through the bed, the contaminant starts to leak out of the system. The relationship of the ratio of the effluent contaminant concentration to the influent concentration verses time or volume can be represented graphically. Initially, the wave front is located near the influent end of the column, and there is complete removal of the contaminant. However, as the mass transfer zone moves through the bed, the contaminant starts leaking. Given enough time, the effluent concentration, C, will attain the same value as the influent concentration. At this point, the adsorber is exhausted and cannot be used further. The carbon is either disposed of or reactivated for reuse.

Practically, the breakpoint defines the exhaustion time for the replacement of carbon. Generally, effluent concentration may not exceed the concentration at the breakpoint because this value is usually set by the regulatory community. Therefore, it is at this point that the operation should be terminated. The exhausted carbon, which has now been contaminated with the contaminants, may be handled by the procedure discussed in the next subsection.

The length of the adsorption zone discussed earlier depends on the waste flow rate. In a long contact time (low flow rate), this zone is usually very narrow, and high efficiency is obtained. However, at low contact time (high flow rate), the zone is spread out and efficiency of carbon usage is dramatically reduced. Therefore, as part of the column studies, optimum contact time should be determined. This can easily be done by having about four laboratory fixed-bed columns arranged in series as opposed to using just one column. This type of setup will yield four different contact times with four different breakthrough histories. Based on the results of such experiments, optimum contact time can be determined.

Not only are laboratory column studies useful in determining the optimum contact time, if the operation is terminated at the breakpoint, the carbon is partially used. In other words, the whole column is not saturated; hence, carbon usage rate will be high, which results in higher operational cost. To avoid this, more than one fixed-bed adsorber is generally used in various practical applications. For example, if two columns that are arranged in series are used, the first column can be operated to complete exhaustion while the effluent from the second column is still below the breakpoint value. When operated in this fashion, the first column is taken out when the concentration from the second column approaches the breakpoint value. The exhausted carbon in the first column is then taken out to be disposed of or reactivated. A fresh carbon column is then placed in front of the second column. Therefore, column studies will indicate the time required for the column to reach exhaustion, and this information can be used to determine the carbon usage.

Having obtained all the laboratory results, the design of a fixed-bed column can then be approached by any of the methods detailed in the literature.[11, 21–23] In addition, mathematical models have been developed to facilitate both feasibility studies and column design. [24]

Carbon Regeneration

The next important evaluation is to determine what to do with the carbon after it is saturated with contaminants. There are generally two options, as indicated in Figure 7.6. One option is to dispose of it in a landfill. This may not be the best option, since there is a potential for water, land, and air pollution. Industries are beginning to revaluate this option because of new regulations and potential liabilities. The next option is the regeneration or reactivation of the carbon. Exhausted carbon may be regenerated with the contaminants recovered for recycling, and the carbon for reuse. On the other hand, the carbon may be thermally reactivated with the destruction of the organic contaminants. The waste quality does not suggest that reclamation of the contaminants is desirable; therefore contaminated carbon should be thermally reactivated. The only concern will be gaseous emissions. In most instances, these contaminants are reduced to carbon dioxide, ash, and other non-toxic products during the reactivation process.

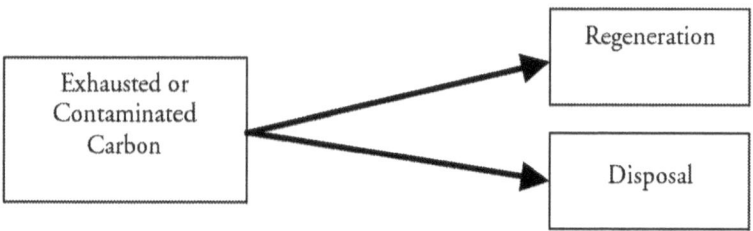

Figure 7.6   Handling of Exhausted Carbon

Evaluation of the reactivation potential is usually conducted in a research kiln in which both the temperature and atmosphere can be controlled. This evaluation determines the quality of the reactivated carbon, the time required, and the amount of carbon lost as a result of such a process. The results should indicate possible emission problems and optimum process conditions required to avoid such problems. Carbon lost during each reactivation process will define the makeup carbon required, which is directly linked to the operational cost.

Other Considerations

Other considerations include contacting systems and modes of operation. Most adsorbers are made of carbon steel or stainless steel. The three contacting systems are: the packed moving bed, the up-flow expanded bed, and the down-flow fixed bed. The down-flow fixed-bed type is most popular.

Down-flow adsorbers are simple to operate, and offer the additional advantage of acting as filters, thus removing suspended solids. In a series arrangement, waste enters the first column with the effluent directed to the second column. When the first column is exhausted, it is taken out to be reactivated. A new column is placed in front of the second column. It is generally more economical to use more than one adsorber for the efficient use of carbon. The operation continues with two columns at all times. In this type of contacting arrangement, a 50%-free board is generally provided for the expansion of carbon during backwashing operations. The waste-flow rate is usually less than or equal to 10 gpm/ft^2, contact time is between 30 and 60 minutes, and the backwash rate is between 15 and 20 gpm/ft^2.

A parallel contacting system may also be used. This type of arrangement is suitable when the waste-flow rate is high enough to require an adsorber with dimensions too large to be economically attractive. When the influent-suspended solids present in the waste are not a concern, an up-flow expanded column arrangement may be used.

Other considerations involve carbon-handling and the reactivation facility. Activated carbon is usually shipped dry, but is usually transported and stored in slurry form. Therefore, provisions should be made for storing and handling. Exhausted carbon can be reactivated on-site or off-site in a suitable furnace. Multiple hearth and rotary-kiln furnaces are generally used. Which of these to use depends on personal preference, economics, and space limitation. Particulate emissions from furnaces can be controlled by the use of scrubbers.

7.8   Ion Exchange

7.8.1   Application

Ion exchange is a process in which ions held by electrostatic forces to charged functional groups on the surface of a solid are exchanged for ions of similar charge in a solution in which the solid is immersed.[11] For years, ion-exchange technology has been used in water-softening, water de-ionization, and industrial and municipal waste treatment. In water softening, ion exchange is used to remove hardness caused by calcium carbonate and magnesium carbonate, and for the removal of iron and manganese from groundwater. In hydrometallurgical practice, ion exchange is mostly applied in the purification and concentration of metals from aqueous solutions.[25, 26] Another application is the removal or recovery of radioactive materials from hospitals, nuclear reactors, and laboratory waste. In most applications, a fixed bed of ion exchange resins is employed.

The four major categories of ion-exchange resins are:[27] strongly acidic cation exchange resins: weakly acidic cation exchange resins; strongly basic anion exchange resins; and weakly basic anion exchange resins. These ion exchangers consist of insoluble polymers to which acid groups or amine groups are chemically attached. The acid groups contain exchangeable cations, and the amine groups contain exchangeable anions. Carriers of exchangeable cations are called cation exchangers, and carriers of exchangeable anions are called anion exchangers. In some cases, the exchangeable group is capable of both cation and anion exchange; hence, the term amphoteric is used to describe the ion exchanger. Liquid ion exchangers are also used for ion exchange with aqueous electrolytes, and for liquid-liquid extraction of electrolytes from aqueous solutions.

Ion exchange operation may be carried out either in batch mode or continuously. In batch mode, waste is mixed with resin in a reactor tank. When the resin becomes saturated with the ions to be removed, the operation is terminated. The saturated resin is separated from the purified water by settling. The resin is then regenerated and the operation continues.

Although ion-exchange operations may be carried out in batch mode, most of the applications use column operation. In this type of operation, waste is passed through a column until breakthrough occurs (the concentration of the effluent reaches the desired level). The ion exchanger is then taken out of service. The saturated ion exchanger can be used again. However, prior to reuse, the column must be backwashed to remove any accumulated dirt. The ion exchanger is subsequently regenerated by passing a concentrated solution containing the exchangeable cation or anion through the bed. This is followed by a rinse operation to remove excess regenerate from the column. The ion-exchange column can now be put back to service.

In an operation in which the objective is to remove both cations and anions, one column removes cations, while another column removes anions.

7.8.2   Theory

When an ion exchanger (in A ionic form) is placed in an electrolyte solution containing counter-ions (B ions), and if the exchanger has greater affinity for the B ions, it will exchange the A ions for the B ions—as shown in equation 7.3—until equilibrium is attained between the solution phase and the resin phase.

$$R^-A^+ + B^+X^- = R^-B^+ + A^+ + x^- \quad \text{(Equation 7.3)}$$

Where:
R = the polymer or matrix containing the permanent group
A = the exchangeable cation leaving the resin
B = the exchangeable cation entering the resin
X = anions in the aqueous phase

At equilibrium, both the exchanger and the solution will contain competing counter-ion species. The concentration ratio of the two competing ions in the resin phase is usually different from that of the solution. The preference of the ion exchanger for one of the counter-ions is often expressed by the separation factor. This quantity is particularly convenient for practical applications such as calculations of column performance. The separation factor (SF) in the above example is defined as:

$$SF = Y_B X_A / Y_A X_B \quad \text{(Equation 7.4)}$$

Where:
Y_A = resin phase equivalent fraction of counter-ion A
Y_B = resin phase equivalent fraction of counter-ion B
X_A = aqueous phase equivalent fraction of counter-ion A
X_B = aqueous phase equivalent fraction of counter-ion B

If ion A is preferred in the resin phase, the separation factor is larger than unity, and if counter-ion B is preferred, the separation factor is smaller than unity. Several factors have been recognized to influence the separation factor or the selectivity in the ion exchange for various cations. These factors include the valence of the counter-ion, ionic solvation and swelling pressure inside the resin phase, and sieve action.

As a rule, the ion exchanger prefers the counter-ion of higher valence. This preference increases with the dilution of the solution, and is strongest with ion exchangers of high internal molality. This can be explained in terms of Donnan potential. Donnan potential attracts counter-ions into the ion exchanger, and thus balances their tendency to diffuse out into the solution.[28] The force with which the potential acts on an ion is proportional to its ionic charge. Hence, the counter-ion of higher charge is more strongly attracted and is preferred by the ion exchanger.

When the resin swells, its elastic matrix is stretched, but because of its elasticity the matrix tends to relax or to contract. It can do so by exchanging a larger counter-ion for a smaller one (on a solvated equivalent volume basis), which causes less swelling. Thus, the ion exchanger prefers the counter-ion with the smaller solvated equivalent volume. Consequently, selectivity increases as swelling pressure increases with dilution of the solution, with a decrease of the equivalent fraction of the smaller ion, and with an increasing degree of resin cross-linking.

Very large organic ions and inorganic complexes may be mechanically excluded from the resin phase by sieve action. Such exclusion occurs if the pores of the resin are too narrow for accommodating the ion.

Similarly, when an ion exchanger is brought into contact with an aqueous medium containing many different counter-ions of different charge—or even of the same charge—the counter-ions will be exchanged for the counter-ion in the ion-exchange resin. The relative affinity of the various counter-ions to the ion exchanger is termed selectivity. The selectivity of an ion towards an ion-exchange resin depends on the aqueous characteristics such as the solution pH, ionic strength, and the presence of competing counter-ions. For example, the selectivity of chelex 100 (an ion exchanger) with various cations in nitrate or chloride solutions is reported to be:[29]

$$Cu^{2+}>>Pb^{2+}>Al^{+3}>Cr^{+3}>Ni^{2+}>Zn^{2+}>Ag^{1+}>Co^{2+}>$$

$$Cd^{2+}>Fe^{2+}>Mn^{2+}>Ba^{2+}>Ca^{2+}>>>Na^{2+}$$

In an acetate buffer solution and at pH 5, the selectivity series for cations is reported to be :[29]

$$Pb^{2+}>Cu^{2+}>>Fe^{2+}>Ni^{2+}>Mn^{2+}>>Ca^{2+}>>Mg^{2+}>>>Na^{1+}$$

These examples of selectivity series readily indicate that the degree of affinity of any cation or anion to an ion exchanger depends not only on the nature of the counter-ion, but also on solution characteristics. It should be noted, however, that the selectivity series for the above cations may be different for a different type of resin.

7.8.3   Design Consideration

Ion exchange operation is essentially an adsorption operation; hence most of the discussions earlier presented for adsorption process are relevant. The activities involved in the selection, and implementation of ion exchange operation are:

- Waste Characterization
- Selection of candidate ion exchangers
- Feasibility studies
- Engineering design

7.8.4   Waste Characterization

The objective of waste characterization is essentially to determine those parameters of the waste that will allow the design engineer to determine whether ion exchange is applicable. This characterization also indicates whether a pretreatment is required, and where an ion exchange column may be placed. Therefore, the following parameters should be determined at minimum:

- Identification and concentration of all organic constituents
- Identification and concentration of all inorganic constituents
- pH
- Suspended solids
- Dissolved Solids
- Oil and grease

Ion-exchange operations are generally applied for the removal of charged species in solution. Therefore, chemical constituents present must be known to the extent that an adequate feasibility study can be designed. In addition, knowledge of the chemical constituents of a solution permits a preliminary determination of the adequacy of ion exchange as opposed to other methods of achieving the same goal. Where there is a variety of chemical constituents present, the analyses will give insight as to other treatment strategies that must be implemented. For example, where there are many organics present, activated carbon may be used to remove the organics prior to the ion exchange. In such a case, the effluent from the carbon will be used for the treatability study, as opposed to using the raw waste.

As previously indicated, pH of the waste affects the ion exchange operation. Not only does it affect the selectivity series, it does affect the optimum capacity of the cations or anions in the ion exchanger. Therefore, an accurate determination of pH is essential. Also, variation in the pH values should be noted. The feasibility study should indicate whether the pH of the various effluents should be adjusted prior to the exchanger. Where it has been predetermined that all aqueous effluents will be combined prior to the exchanger, the pH of the combined stream is required.

One of the common problems encountered in the use of ion exchangers is clogging. Therefore, the determination of solids-concentration is essential. The result may indicate that a pretreatment process such as filtration is required. Oil and grease concentration should also be noted, since their presence in high concentration may disrupt the operation. Once again, where high concentration is encountered, pretreatment is required.

7.8.4   Feasibility Studies

Having discovered the characteristics of the waste, the next logical step is to determine the parameters for the design. However, prior to this step it is not yet clear what type of ion exchange to select. One can usually approach this question by obtaining as much information as possible. Sources of information include ion exchange manufacturers, consultants, and other institutions that use ion exchange treatment. Other industries that produce similar waste and use ion exchange treatment are usually the most reliable sources. While getting this type of information, it may be beneficial to visit such a facility, and gain some insight into the type of problems they normally run into. With the above information, one is in the position to select a minimum of three candidate ion exchangers for evaluation. Manufacturers may be requested to send some samples for this evaluation. Usually, these samples come with some literature that may guide the feasibility study. This literature generally contains information on the capacity of the various cations and the selectivity series. The feasibility studies are then designed to determine the capacity of the ion exchanger for the cations or anions, the column design parameters, and the regeneration requirements.

Equilibrium Studies

Ion exchangers are characterized in a quantitative manner by their capacity, which in common usage is defined as the amount of ions (in equivalents) sorbed by a specific amount of resin. Other definitions and units are given elsewhere.[28] The capacity can usually be determined in the laboratory by the use of batch reactors. In this process, various amounts of the resin are brought into contact with a known volume and concentration of the cation or anion. At equilibrium, the solution phase is analyzed. From a mass balance, the amount adsorbed in the resin can easily be calculated. Usually the experiment is set up with a series of batch reactors, so that a wide range of concentration is covered. The resulting data may mathematically be treated using similar isotherm or equilibrium models to those discussed in the section on adsorption. Note that while equilibrium may take weeks to be achieved in the case of carbon adsorption, equilibrium in ion-exchange batch operations may be reached in a matter of hours to a few days. The equilibrium capacity so obtained may vary if the solution characteristics is varied. Strong-acid and strong-base groups are by definition completely ionized under experimental conditions, and so their capacities are constant. On the other hand, for weak-acid and weak-base groups, the capacity will strictly depend on the solution pH. Hence, for the efficient utilization of the resin, the solution pH should be more than the pK values of the groups attached to the resin matrix.

Column Studies

Because the fixed-bed mode of operation is commonly used, columnar studies should be conducted with the best candidate resin as determined above to simulate actual operation. The aim here is to determine those parameters that affect the operation and eventually can be used in the actual design. The important parameters that need to be established are:

- contact time
- resin-usage rate
- regeneration requirement

Column studies are conducted in a manner similar to that reported earlier for adsorption operations. This is done by passing waste with a known concentration through the fixed-bed ion exchanger as shown in Figure 7.7. Using our previous example, it is assumed that the resin is in the A ionic form and that ion B is present in the waste. The objective is to remove ion B from the solution. As the waste is passed through the column, ion B is exchanged for ion A. Figure 7.7 may be used to understand events in the ion-exchange column. Plotted for either counter-ion is the ratio of the effluent concentration to the initial concentration versus time or volume of waste passed through the column.

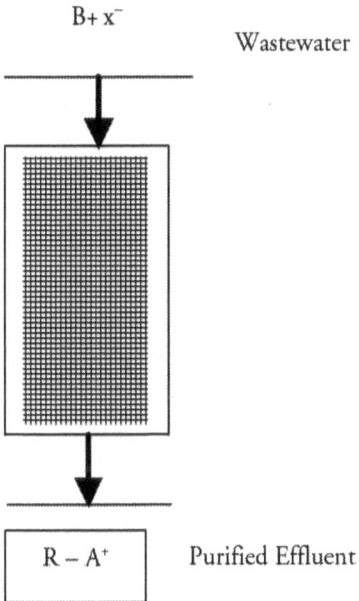

Figure 7.7   Schematic of Ion Exchange Fixed Bed

The two curves represent the effluent histories or breakthrough profiles of the two ions. Factors that affect the breakthrough profile include the ion-exchange capacity of the ion exchanger, its ionic form, the composition of the solution, the equilibrium relationship between the solution and the exchanger, the rate of ion exchange, and the superficial velocity.[30] Initially in the process, no ion B will be present in the effluent stream, but because it is being adsorbed onto the exchanger with the desorption of ion A, ion A is present in the effluent. As the operation progresses, and as the ion exchange becomes partially saturated with ion B, ion B will then start to leak out of the column. When the effluent concentration of ion B reaches the permissible level (the breakthrough point), the column is said to be exhausted and the operation must be terminated. The resin is then backwashed to remove any clogging material. The column is subsequently regenerated and rinsed, and the operation may be continued. It is noted that if the operation is carried out in this fashion, the ion exchange capacity is not efficiently utilized. Therefore, several columns may be operated in series. When this is done, the first column may be operated to maximum capacity without the concentration of ion B in the lead column exceeding the permissible level. Series operation or parallel modes have earlier been discussed in the section on adsorption. The above experiments may be repeated for various waste-flow rates from which optimum flow-rate, and hence the contact time, will be determined,. Resin usage rate can be calculated at this point.

When the ion exchange becomes saturated, it must be regenerated prior to reuse. The regeneration operation may be represented by the following equation:

$$R^-B^+ + A^+ = R^-A^- \quad \text{(Equation 7.5)}$$

The objective of the regeneration operation is to revert the resin to its original form, as indicated in the above equation. Regeneration can be achieved by passing a concentrated solution containing the original ion, which reverses the equilibrium. Note that a column treatability study is required to determine the optimum regenerant concentration needed, as well as the optimum flow rate. Strong acidic resin originally in hydronium form can be regenerated with hydrochloric acid or sulfuric acid. Sodium chloride is the choice when the resin is initially in the sodium form. Sodium hydroxide or ammonium hydroxide can be used to regenerate the common anion ion exchangers. However, the choice of a regenerant depends on the original ionic form of the resin and the associated costs of the regenerants. The effluent from the regeneration process has a high concentration of ion B. This can be reused or discarded, depending on the economic value of the recovered material.

Other Considerations

After the above investigation, the ion-exchange column can easily be designed with any of the methods detailed in the literature.[11, 13] Typical design parameters include:[11] column depth of between 2–6 feet, provision of 50–100% of packed column height for expansion during backwashing, waste flow rate of 5–10 gpm/ft², regenerant flow rate of 1–2 gpm/ft², rinse-water volume of 30–100 g/ft³, and rinse-water flow rate of 1–1.5 gpm/ft².

Although the above numbers may give an idea of design parameters, the best numbers are always obtained through treatability or pilot studies. Another consideration is the material used in the construction of the column. Since both acids and alkaline substances will be used in the regeneration steps, the material has to be such that there is no degradation. Therefore, the internal walls may be coated with high-resistance materials.

Economic Considerations

The ion-exchange process, like other unit processes, requires detailed economic evaluation prior to implementation. This evaluation involves the analyses of both capital and operation costs. Capital costs may be high depending on the type of material used for the construction of the columns. These columns must be able to withstand extreme pH values. The main operation cost has to do with the regenerants and resin replacement. The cost generally varies in relation to the quality of the waste.

7.9 Air-Stripping

7.9.1 Application

It was earlier presented in the adsorption section that carbon adsorption can be used for the removal of toxic organic compounds from water or waste. One of the problems commonly encountered in carbon adsorption operations is that some classes of organics are not effectively removed. These classes of organics break through prematurely, and thus render the carbon column ineffective for the purification process. Common characteristics of these compounds include high vapor pressure and low solubility in water. These organic compounds are generally referred to as the volatile organic compounds. Air-stripping is one treatment technology that can be used to remove these volatile compounds from water or waste. The air-stripping operation has been used for years, and is presently being widely applied in groundwater remediation programs for the treatment of groundwater contaminated by volatile organic compounds. This treatment technology has advantages such as simplicity of operation, low capital cost, and very low operation and maintenance costs when compared to other methods of volatile organic removal. [32]

There are various operational contacting modes or equipment configurations available for air-stripping technology. The major ones include countercurrent packed columns, diffused aeration, cross-flow towers, and tray aerators. In the countercurrent packed column operation, the water or waste containing the contaminants is allowed to flow down the packed column, with air flowing in the opposite direction. The packing material provides intimate contact between the two phases, and thus enhances the transfer of the contaminants from the water phase into the air phase. In the diffused-air operation, the water flows down into the diffused-air basin, with the air from the diffusers located at the bottom of the basin rising. The cross-flow operation is similar to the countercurrent operation. The main difference is that as opposed to the countercurrent operation, air is pulled across the packed column. In the trayed aeration system, contaminated water trickles down through series of trays as air flows up the column.

Although different contacting systems exist, most applications use the countercurrent packed-bed system because it is the most efficient. Irrespective of the contacting mode, it should be noted that this operation transfers contaminants from the aqueous phase into air. Recent regulations suggest that discharge of these contaminants into the atmosphere is not desirable. The countercurrent equipment can easily be connected to vapor purification equipment, which is an additional advantage.

7.9.2 Theory

The strippability of a volatile organic compound depends strictly on its ability to transfer from the aqueous phase into the gaseous phase. This is a mass transfer operation whose detailed theory is well documented in various chemical engineering texts.[31, 33, 34]

When waste containing volatile organics is exposed to air, the organics will tend to leave the aqueous phase and enter the gaseous phase. The amount of material and the rate of mass transferred depends on the concentration gradient, among other things. The rate of mass transfer can mathematically be expressed as follows:

$$M = Ka\,(C_1 - C_g) \quad \text{(Equation 7.6)}$$

Where:

M = rate of mass transfer

K = mass transfer coefficient

a = effective area

C_1 = contaminant concentration in the aqueous phase

C_g = contaminant concentration in the gaseous phase

$(C_1 - C_g)$ = driving force

Equation 7.6 shows that the rate of mass transfer is a function of the mass-transfer coefficient, effective area, and the driving force. Due to experimental difficulties in the determination of "K" and "a" separately, they are usually determined as one parameter (Ka), which is generally termed the mass overall transfer coefficient. This coefficient is a function of several factors, including the packing material, airflow rate, water-flow rate, and temperature. The driving force determines the amount of material to be transferred from the aqueous phase into the gaseous phase. The higher the driving force, the higher the rate of mass transfer. The driving force also relates the condition within the air-stripper to the maximum condition obtainable. This maximum condition is the equilibrium condition. In mathematical modeling or analyses of air-stripping operations, it is generally assumed that equilibrium exists at the liquid-gas interface. Therefore, the concentration at the interface is governed by Henry's law. Henry's law can mathematically be represented as:

$$C_{eg} = HC_{el} \quad \text{(Equation 7.7)}$$

Where:

C_{eg} = equilibrium concentration in the gas phase

C_{el} = equilibrium concentration in the liquid phase

H = Henry's law constant

The above equation simply states that the concentration in the gas phase is the product of the concentration in the liquid phase and proportionally constant (Henry's-law constant).

Henry's-law constant is a property of the compound, and its value can be used to determine whether the compound can be removed from the liquid phase by air-stripping. In general, a compound with a higher Henry's-law constant is more strippable than one with a lower value. For example, among the volatile organics present in electronics-manufacturing waste are chloroform, trichloroethylene, and toluene. Their respective dimensionless Henry's-law constants at 20°C are 0.14,

0.42, and 0.24.[35] These values indicate that trichloroethylene is more strippable than either chloroform or toluene, and that toluene is more strippable than chloroform. In general, compounds with Henry's-law constants above 0.1 are readily amenable to air-stripping, and those with constants of 0.02 to 0.1 are strippable with difficulty, requiring packing depth or air-to-water ratio.[36] The Henry's-law constant can easily be determined experimentally. On the other hand, it can also be calculated with the following expression:

$$H = P/C_s \quad \text{(Equation 7.8)}$$

Where:
P = vapor pressure of the compound
C_s = solubility of the compound

The above expression indicates that the Henry's-law constant of a given compound can easily be estimated through knowing both vapor pressure and water solubility. Fortunately, these values are readily available from various handbooks such as the Handbook of Chemistry and Physics.[37] It should be noted that both solubility and vapor pressure are temperature-sensitive, and hence the Henry's-law constant is also temperature sensitive.

7.9.3   Design Considerations

Earlier it was stated that the packed countercurrent air-stripping operation is the most efficient mode of operation. Therefore, the rest of this section will focus on this type of contacting system. Figure 7.8 shows the schematics of such an operation. Contaminated water flows down the packed column, with uncontaminated air flowing in the opposite direction. As a result of mass transfer from the aqueous phase into the gaseous phase, the contaminants are transferred into the gaseous phase. The air exiting from the top of the column is thus contaminated, while the water leaving the column from the bottom has been purified. For such a process, it is essential to understand the parameters that govern the degree of removability of the contaminants. This type of understanding can be gained through various types of studies, including waste-characterization and feasibility studies.

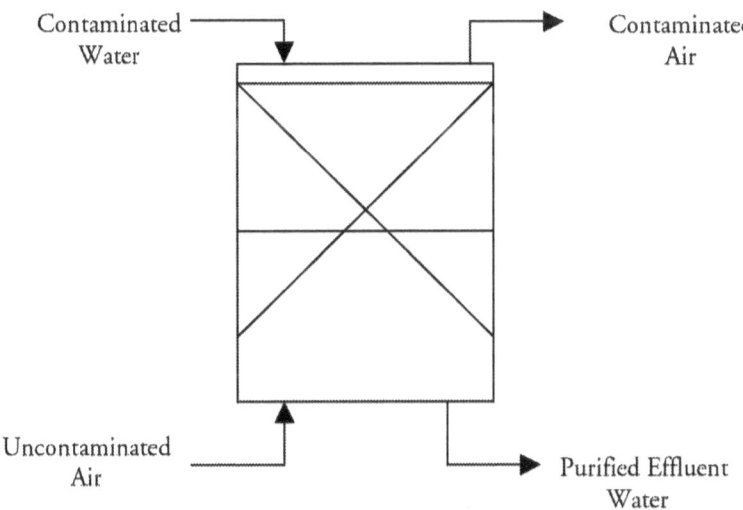

Figure 7.8   Schematic of Air-Stripping Operation

7.9.4   Waste Characterization

Knowledge of waste characteristics is essential for the implementation of any treatment technology. For the most part, one has to know the concentrations of the substances present, determine the applicable technologies, and determine whether those technologies can achieve the treatment goal. Waste characterization also indicates the need for pretreatment. At a minimum, the following should be determined:

• identity and concentration of all organic constituents
• identity and concentration of all inorganic constituents
• pH
• suspended solids
• dissolved solids
• oil and grease

The waste should be analyzed for all organics and inorganics present. Since air-stripping mainly removes volatile organics, the presence of other organics with high solubilities and low vapor pressures—and hence low Henry's-law constants—will indicate that another treatment technology may have to be used in combination with air-stripping. In this regard, carbon adsorption may precede air-stripping. The advantage of such an arrangement is that the carbon will remove some of the volatile compounds, and hence decrease the mass loading on the stripper. Concentration of the organic compounds will permit adequate planning of the feasibility study. The presence of various inorganics at high concentrations indicate that pretreatment should occur prior to air-stripping. Precipitation or ion-exchange technologies may be applied. Therefore, the nature and concentration of the various organics and inorganics present is the most important fundamental information required. In addition, the waste-flow rate should be determined, and any fluctuations should be noted. Where waste-flow rate fluctuates, a detention tank may be used prior to the air-stripping, since a steady-state condition is desirable.

The pH of waste should be determined, and variations within the various sources noted. The pH of an aqueous system generally governs the state of the chemical species present. In addition, extreme pH values may result in the deterioration of the packing material, and may thus affect the removal process. Where processes such as precipitation precede air-stripping, pH adjustment may be required prior to stripping.

The purpose of packing material is to provide maximum contact between the aqueous phase and the gaseous phase, thereby enhancing mass transfer. These packing materials are analogous to the carbon or ion exchange materials presented in their respective sections. It has been noted that a common problem in those operations is clogging. Therefore, it is important to determine the concentration of solids present, and hence determine whether pretreatment is required. If necessary, filtration may precede air-stripping.

Oil and grease presence should also be determined. If they are present at high concentrations, removal is required prior to air-stripping.

7.9.5   Feasibility Study

From the feasibility study, design parameters may be established. By this point, pretreatment requirements should be known and should constitute part of the feasibility study evaluation. The effect of the various design parameters on the treatment efficiency should also be evaluated. For this type of application, there are two types of feasibility study:

- equilibrium studies
- column studies

Equilibrium Studies

The main objective of equilibrium studies is to determine the Henry's-law constant. As earlier mentioned, this parameter can easily be calculated by knowing both the vapor pressure and the water solubility of the compound. It can also be determined experimentally. This can be done by placing various volumes of waste in various reactor bottles, and allowing equilibrium to be attained between the waste and the void space. The reactor bottle should be agitated periodically to maintain complete mix condition. At equilibrium, the liquid-phase concentration is then analyzed. The corresponding gas-phase concentration can then be calculated through mass balance. When the gaseous-phase equilibrium concentration is plotted against the aqueous-phase equilibrium concentration (see equation 6.7) on a graph, a straight line is usually obtained. The Henry's-law constant is the slope of such a line. A regression analysis is usually performed to obtain the best value for the constant.

Column Studies

Prior to a column study, packing material should be selected. Various types of packing material—such as Pall Rings and Super Intalox—are available. These packing materials come in various sizes (e.g., 1", 2"). Selection of size and shape involves a compromise between low packing factor—hence minimum head loss (minimum tower diameter)—and maximum overall rate of transfer.[38] Manufacturers of these packing materials usually furnish information regarding the characteristics of the packing material. This information, plus information from other institutions where this technology is applied, can be very useful in the selection process.

Through column studies, the optimum air-to-water ratio can be determined. During a normal air-stripping operation, waste flows down the column while air flows in the opposite direction. However, if the water-flow rate is held constant, and the air-flow rate is increased, a point is reached at which entrainment of the liquid by the rising air may occur, which is characterized by a sudden pressure drop.[38] This condition, generally referred as to flooding, is to be avoided. A rule of thumb used for establishing air-velocity is that an acceptable air-velocity is 60% of the air-velocity at flooding.[39] Data resulting from the column experiment can also be used to calculate the overall mass transfer coefficient, as detailed elsewhere.[33, 34, 38] This coefficient depends on the geometry of the tower, the air-to-water ratio, the type of packing material, the temperature, and the type of contaminant being removed.

Other Considerations

At the termination of the feasibility study, the engineer is equipped with the design parameters. For example, after the determination of the optimum air-to-water ratio, the cross-sectional area of the column can be calculated by dividing the airflow rate by the air velocity. Since other parameters such as influent concentration, desired effluent concentration, Henry's-law constant, and the overall mass transfer coefficient have already been established, the height of packing can be calculated as detailed elsewhere.[31, 33, 34, 38] In some cases, the calculated column size may be economically unattractive or impractical; hence, two or more columns in series may be appropriate to achieve the desired effluent level. Other components of the air-stripping operation are the water-distribution system at the top of the column, the air blower, and the collection sump at the bottom of the column. Once a detailed design is developed, these components can then be selected.

Design parameters vary depending on the waste characteristics and contaminants of concern. However, an air-to-water ratio of around 50:1 (volume basis), and pressure-drops of about 0.25 inches water/ft packing depth will provide a good design with flexibility for water-flow variations.[36]

Although the above discussion assumes waste with only one major contaminant of concern, it must be stressed that the calculations are still valid for a multi-component system, except that more extensive calculations are required. Where waste is contaminated with many volatile organic compounds, the optimum air-to-water ratio is computed for each of the contaminants. The highest air-to-water ratio is then used for the column design. Similarly, the depth of packing should be calculated for each compound, with the highest value being the deciding factor.

As mentioned earlier, it is essential to select construction materials that are highly resistant to corrosion. This is particularly useful when the pH of the waste varies with no pretreatment. Common problems encountered in air-stripping operations include precipitation of various metallic oxides onto the packing material, or even biological growth. These problems can be corrected by washing the column with acidic or alkaline solutions.

Air-stripping operations transfer contaminants from one phase into another phase, and in this case from water to air. Therefore, to avoid air-pollution problems, vapor purification is essential. Activated-carbon technology can be used for this purpose. The system can be designed as previously indicated. Equilibrium studies and other feasibility studies should be undertaken to optimize such a system. The contaminated carbon can then be regenerated for reuse with destruction of the organic compounds. Where contaminant recovery is desired, steam regeneration can be employed. In such an operation, the condensate, which is now highly concentrated, can either be reused or disposed. Because of possible future financial liability associated with land disposal, thermal regeneration of the exhausted carbon is preferred.

Economic Consideration

Major cost items for air-stripping include the stripper column, the water distribution column at the top of the column, the air blower, and the collection sump at the bottom of the column. Cost-affecting factors include waste characteristics that affect the choice of construction materials, water-flow rates that control the design of the distribution system, and airflow rates that affect the size of the air blower. Other cost items include the packing materials, pumps, and valves. To a large extent, the operational costs are affected by vapor recovery, precipitation and biological growth on the packing material, and packing-material replacement. Therefore, the capital and operation costs for air-stripping generally depend on the size of the system. Fortunately, these costs compare favorably with the available alternatives.

7.10 Deep Well Re-injection

7.10.1 Applicability of Technology

Re-injection is a useful method of fluid waste disposal. This technology, also referred to as deep well injection or subsurface injection, utilizes deep subsurface formations to trap waste within the ground. Injection wells are used to transport the waste below-ground, where it is stored and sometimes treated. The United States Environmental Protection Agency (EPA) has classified the different varieties of contemporary injection wells according to the types of waste they inject,[1] as follows:

- Class I—Technologically advanced wells used to inject hazardous and non-hazardous wastes.

- Class II—Wells used to inject fluids associated with oil and gas production, specifically brine.

- Class III—Wells used to inject superheated steam or water for mineral extraction operations.

- Class IV—Wells used to inject hazardous or radioactive waste into or above underground sources of drinking water. Class IV injection wells are banned by the EPA.
- Class V—All other wells not fitting other classifications, generally low-tech equipment.

Injection wells came to prominence in the 1930s in the oil industry. Injection well use has continued to expand, as re-injection technology is now utilized in a variety of waste management applications.

Table 7.14   Contaminants Compatible with Re-injection Management

- Hydrocarbon products (generally for storage, not treatment)
- Brine (oil extraction byproduct)
- Volatile organic compounds
- Semi-volatile organic compounds
- Fuels
- Explosives
- Pesticides

7.10.2 Description

A re-injection application consists of several process components. Typically, re-injection entails site evaluation, well construction, well operation, monitoring, and process evaluation. Despite the variability of injection well uses, the procedure for implementing an application is uniform for the five classes of wells. The re-injection process components are depicted in Figure 7.9.

A thorough site evaluation precedes injection well construction. The investigation of subsurface geology at a potential site is the most crucial phase of the re-injection process, as the geology must be appropriate for the application. Impermeable layers that prevent liquid-waste migration must isolate the confining formation and, ideally, a minimal number of freshwater aquifers will be present.

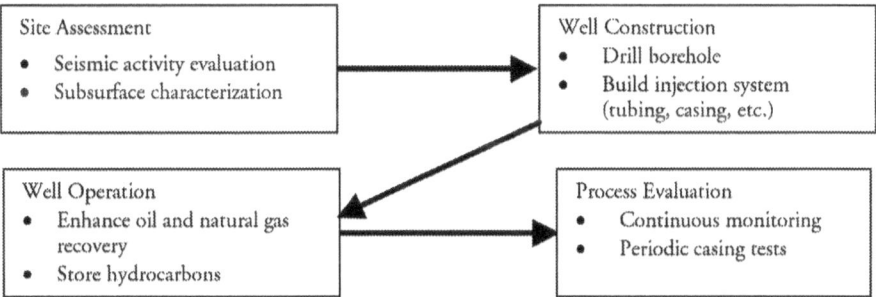

Figure 7.9   Re-injection Process Flow Chart

Well construction follows the site evaluation if conditions are determined to be suitable for re-injection. The first step is to drill the borehole for the injection well. The borehole is drilled into a confining formation that does not contain potential underground sources of drinking water (USDW). Injection-well depths frequently exceed 4,000 feet. Once the borehole is complete, piping infrastructure is installed. A typical injection well consists of three concentric layers of pipe: the outer (surface) casing, the string casing, and the annulus. The outer casing protects the string casing from the actual sediment punctured by the borehole, and is cemented to the surface to prevent contamination. The string casing is filled with cement to prevent leakage of waste into the subsurface and also contains the injection tubing through which the waste is injected. The annulus lies between the string casing and the injection tube itself, and is filled with pressurized fluid in order to prevent the backup of waste.[2]

The well is monitored closely during operation to check for leaks or other deficiencies that could compromise the operation. Continuous monitoring of the well ensures regulatory compliance, high efficiency, and early detection of any problems that could adversely affect the local environment.

7.10.3 Advantages and Disadvantages of Re-injection

The most significant advantage of re-injection is its ability to enhance oil and natural gas production. Re-injection during oil extraction is a well-documented way to increase production rates and prolong the life of an oil field. The process, widely known as enhanced oil recovery (EOR), can consist of secondary recovery, tertiary recovery, or a combination of the two. Secondary recovery involves re-injection of brine to expedite oil recovery, whereas tertiary recovery involves re-injection of gases or chemical additives to prolong oil-field life.[3]

Another attractive aspect of re-injection technology is for *in-situ* leaching operations. *In-situ* leaching is a means of mining various metals, including gold and uranium, from the subsurface. An injection well is used to convey a neutral fluid solution into the subsurface. Mineral ores leach into the fluid solution as it circulates through the subsurface. The solution is subsequently brought to the surface and is processed for recovery of the metals.

Equally important is the fact that contact with waste is minimized by the very nature of the injection well. Injection-well operators may be the last living organisms to be in contact with waste for an extended period of time. Once injected, waste is confined thousands of feet below the surface of the earth and is thus rendered inaccessible to humans, plants, or animals.

There are some drawbacks associated with the use of re-injection technology. There are extensive regulations governing the operation of injection wells. Many statutes exist at the federal level, particularly through the EPA's Underground Injection Control (UIC) program, which works with state and local governments to regulate injection well use.[1] Injection well operation is not feasible in seismically active areas, due to the possibility of waste migration along active faults. In some instances, re-injection has been linked to earthquakes. Additionally, re-injection is potentially harmful to underground sources of drinking water (USDW). Although measures are taken to construct re-injection wells at locations where aquifers are not present, leakage into the environment and subsequent contamination of USDW remains a major concern.

7.10.4 Design Considerations

A re-injection application must be carefully planned, accounting for several important factors to ensure proper operation. Perhaps the most important consideration is subsurface geology. Subsurface features must be conducive to well operation, and the presence of aquifers must not be overlooked. Specifically, subsurface formations must be able to isolate waste by preventing migration, and injection wells must be operated in a fashion that does not contaminate local USDWs. A significant amount of information on local sediments must be compiled as well, including suspended-solids content, biological and chemical oxygen demand, total organic carbon, and pH.

Injection wells are subject to an extensive list of regulations. Federal, state, and local laws regulate the siting, construction, and operation of injection wells. States with EPA-approved Underground Injection Control programs have the option of adopting more stringent guidelines for injection-well operation.

Another consideration is the compatibility of injected fluid with the injection well infrastructure. Generally, wastes should be neutralized prior to injection to prevent fouling of the system. An abundance of suspended solids in the fluid waste can plug the injection infrastructure. Excessive amounts of organic carbon can also cause fouling of the system by stimulating microbial growth. Injection wells are closely monitored to detect fouling or leakage. Operators of Class II injection wells must test well casings every five years for leaks.[1] In addition, the results of pressure-monitoring must be presented to the regulatory authorities monthly. The required reports must include details of pressure-monitoring for several parameters, including annulus and injection pressure.

7.10.5 Environmental Effects

The greatest threats to the environment posed by injection-well operation are groundwater contamination and potential geological problems. Improper operation of injection wells can contaminate USDWs, particularly if the wells are operated in close proximity to aquifers. Several sets of regulations apply to all aspects of injection-well operation as previously indicated. Further, it is inappropriate to operate injection wells in seismically active regions. Injection-well operation can promote instability in subsurface formations and has been linked to earthquakes.

7.10.6 Results of Case Studies

7.10.6.1 Romulus, Michigan

Efforts are currently underway by Environmental Disposal Systems, Inc. (EDS) to begin the operation of two deep re-injection wells at a site in Romulus, Michigan. The wells were drilled to a depth of over 4,000 feet and are to be used as Class I wells for the disposal of hazardous wastes including heavy metals, acids, and solvents. EDS, a waste disposal company, had planned to begin operations upon well completion in 1994, but has faced a lengthy delay due to the efforts of local residents to prevent hazardous waste re-injection from occurring in their community.[4] Increasing public concern regarding activities with the potential for environmental impact, as well as knowledge of adverse health effects from deep well re-injection operations in Texas and Oklahoma, have galvanized the community to object to the hazardous waste disposal operations proposed by EDS in Romulus.[5]

The EPA considers deep well re-injection to be a safe and effective method of hazardous waste disposal, and has granted operational permits for approximately 163 Class I wells similar to the EDS wells, at 51 locations throughout the United States.[6] Based on compliance-study information provided by EDS, its wells have been situated in an ideal geological formation: a reservoir of porous rock that will provide containment of the hazardous wastes for a minimum of 10,000 years. This reservoir is surrounded by a confining layer that will prevent wastes from migrating beyond the boundaries of the reservoir.[7] It is estimated that over 7 million gallons of wastes could be injected into the two EDS wells each month.[5, 7]

In December 2002, the EPA initiated the public consultation process regarding its plan to issue a land-disposal restriction exemption to EDS, which would allow it to begin re-injection operations.[7] The EPA granted the land-use exemption to EDS in March 2004.[1] However, EDS must still obtain approval from the Michigan Department of Environmental Quality in order to operate the wells.[8] As of August 2004, EDS still had not begun to operate the two deep re-injection wells in Romulus, Michigan.

7.10.7 Engineering Economics

When he compared it to other major waste-disposal technologies such as landfilling and incineration, Harding (2003) determined that deep-well injection was the most effective method of waste disposal in terms of costs to construct, operate, and close.[6] These findings were also used to support Harding's opinion that deep-well injection is an invaluable technology for waste disposal, as it does not discharge into sewer systems and is constructed to prohibit releases into the drinking-water supply. However, other research has focused on the danger of, and the lack of demand for, this method of waste disposal, due to the potential for leakage into subsurface aquifers and a growing movement in industry to eliminate waste production at the source, thus reducing the amount and capacity of waste-disposal facilities required.[5, 8]

Table 7.15   Comparison of Waste Treatment and Disposal Technologies

	Waste Disposal Technology		
	Deep Well Injection	Landfilling	Incineration
Construction Costs (millions of dollars)	3	2 to 7	300
Annual O&M Costs (millions of dollars)	4.4	0.48 to 1.68	36

	Waste Disposal Technology		
	Deep Well Injection	**Landfilling**	**Incineration**
Life of Design (years)	20	15	20
Proximity to Water Sources	Isolated from Drinking Water	Direct	Indirect
Discharges per Year per Unit to Surface Waters (gallons)	None	Leachate, methane, hexane	CO_2, ash, metals, dioxins
Disposal Costs (dollars per gallon)	0.25 to 1.50	0.39 to 0.78	0.8 to 48
Average Annual Volume Treated per Unit (millions of dollars)	10	20.25	2
Total Cost ÷ Average Annual Volume	0.78	0.14 to 0.50	169.50

ENDNOTES

Oil Industries Fundamental

1. Royal Dutch Shell Group Staff. 1983. The Petroleum Handbook, 6th Edition. Amsterdam: Elsevier Science. 710

2. Yergin, Daniel. 1999. The Prize.

3. Pees, Samuel T. 2003. Oil History. www.oilhistory.com

4. Gaston, Robert. 2001. Discovery of the Spindletop Oilfield, www.drillinginfo.com/wonderings20000101.html

5. Dyk, Karl. 1961. Geophysical Surveys. Chapter 11. In: Moody, Graham B. Petroleum Exploration Handbook. New York: McGraw-Hill.

6. Jahn, F., Cook, M., and Graham, M. 1998. Hydrocarbon Exploration and Production. Developments in Petroleum Science, Vol. 46. New York: Elsevier Science. 384

7. Von Buelow, EU Essential Points in Use of Geophysics. Oil & Gas Journal, Special Issue: 100 Years of Industry Leadership. Vol. 100, Iss. 35, Aug. 2002. Pages 56–66.

8. Bott, Robert. 1999. Exploring Canada's Oil-and-gas industry, 6th Edition. Calgary, Alberta: Petroleum Communication Foundation. 101

9. North, F. K. 1985. Petroleum Geology. Winchester, Massachusetts: Allen & Unwin Inc. 607

10. Freudenrich, Craig C. 2003. How Oil Drilling Works. www.howstuffworks.com/oil-drilling.htm

11. Natural Gas Supply Association, The. 2003. www.naturalgas.org

12. Bell, H. S. 1959. American Petroleum Refining, 4th Edition. Princeton, New Jersey: Van Nostrand Company. 538

13. Freudenrich, Craig C. 2003. How Oil Refining Works. www.howstuffworks.com/oil-refining.htm

14. Speight, James G. 2002. Handbook of Petroleum Analysis. Hoboken, New Jersey: John Wiley & Sons Inc. 409

15. Rodrigue, Jean-Paul. 2004. Transport Geography on the Web. Hofstra University Dept. of Economics and Geography. http://people.hofstra.edu/geotrans

Waste Characteristics

1. United States Department of Energy, Office of Fossil Energy. 1999. Environmental Benefits of Advanced Oil and Gas Exploration and Production Technology. Washington.

2. American Petroleum Institute. 1997. Environmental Guidance Document: Waste Management In Exploration and Production Operations, 2nd Edition. Washington DC: American Petroleum Institute. 78.

3. Bott, Robert 1999. Exploring Canada's Oil-and-gas industry, 6th Edition. Calgary, Alberta: Petroleum Communication Foundation. 101.

4. Jahn, F., M. Cook and M. Graham, 1998. Hydrocarbon Exploration and Production. Developments in Petroleum Science, Vol. 46. New York: Elsevier Science. 384.

5. Royal Dutch Shell Group Staff. The Petroleum Handbook, 6th Edition. 1983. Amsterdam: Elsevier Science. 710.

6. ICF Consulting. 2000. Overview of Exploration and Production Waste Volumes and Waste Management Practices in the United States. American Petroleum Institute. 112.

7. Patin, Stanislav (translation by Elena Cascio). 1999. Environmental Impact of the Offshore Oil-and-gas industry. Chapter: Waste Discharges. www.offshore-environment.com/discharges.html.

8. Wills, Jonathan. 2000. Muddied Waters: A Survey of Offshore Oilfield Drilling Wastes and Disposal Techniques to Reduce the Ecological Impact of Sea Dumping. Sakhalin Environment Watch. http://www.offshore-environment.com/jonathanwills.html

9. Smith, K. P. 1992. An Overview of Naturally Occurring Radioactive Materials (NORM) in the Petroleum Industry. United States Department of Energy,

Office of Domestic and International Energy Policy. Argonne, Illinois: Argonne National Laboratory.

10. Sittig, Marshall. 1978. Petroleum Refining Industry: Energy Saving and Environmental Control. Energy Technology Review No. 24, Pollution Technology Review No. 39. Park Ridge, New Jersey: Noyes Data Corporation. 374.

11. Manahan, Stanley E. 1997. Environmental Science and Technology. Boca Raton, Florida: Lewis Publishers. 641.

12. Speight, James G. 2002. Handbook of Petroleum Analysis. Hoboken, New Jersey: John Wiley & Sons Inc. 409.

Environmental Issues of Oil Industry

1. Jones, Laura, Fredericksen, Liv and Tracy Wates. 2002. Critical Issues Bulletin: Environmental Indicators, 5th Edition. Vancouver: Fraser Institute. 136.

2. Fay, James A. and Dan S. Golomb 2002. Energy and the Environment. New York: Oxford University Press. 314.

3. United States Environmental Protection Agency. Air Quality: Six Common Air Pollutants. http://www.epa.gov/air/urbanair/6poll.html

4. United States Environmental Protection Agency. Clean Air Markets: Environmental Issues. http://www.epa.gov/airmarkets/envissues/index.html

5. Manahan, Stanley E. 1997. Environmental Science and Technology. Boca Raton, Florida: Lewis Publishers. 641.

6. United States Environmental Protection Agency, Office of Air Quality Planning and Standards. 1998. NOx: How Nitrogen Oxides Affect the Way We Live and Breathe. Research Triangle Park, North Carolina.

7. Bott, Robert 1999. Exploring Canada's Oil-and-gas industry, 6th Edition. Calgary, Alberta: Petroleum Communication Foundation. 101

Environmental Regulations

1. American Petroleum Institute. 1997. Environmental Guidance Document: Waste Management In Exploration and Production Operations, 2nd Edition. Washington, DC: American Petroleum Institute. 78.

2. United Kingdom Department of Trade and Industry. 2003. Environmental Legislation Applicable to the Onshore Oil Industry in England, Scotland, and Wales.

www.og.dti.gov.uk/regulation/legislation/environment/
onshore_hydrocarbons/index.htm

3. National Oceanic and Atmospheric Administration (NOAA), Hazardous Materials Response and Assessment Division. 1992. Summaries of United States and Significant International Spills. Seattle, Washington: United States Department of Commerce. 224.

4. Patin, Stanislav (translation by Elena Cascio). 1999. Environmental Impact of the Offshore Oil-and-gas industry. Chapter: Conventions Regulating Environmental Impact. www.offshore-environment.com/conventions.html.

5. Inho Kim, 2002. Ten years after the enactment of the Oil Pollution Act of 1990: a success or a failure? Marine Policy, v. 26, p. 197–207.

6. Ross, William M. 1973. Oil Pollution as an International Problem: A Study of Puget Sound and the Strait of Georgia. Western Geographical Series, Volume 6. Department of Geography, University of Victoria, British Columbia. 278.

7. Fridtjof Nansen Institute. 2003. Yearbook of International Co-operation on Environment and Development. Agreements on Environment and Development: the Marine Environment. http://www.greenyearbook.org

8. International Maritime Organization. 2003. International Convention for the Prevention of Pollution from Ships, 1973 (MARPOL). www.imo.org/Conventions/mainframe.asp

9. Office for the London Convention. Convention on the Prevention of Marine Pollution by Dumping of Wastes and Other Matter, 1972, http://www.londonconvention.org/main.htm

10. United Nations. Conference on Environment and Development (UNCED), 1992. "Rio Earth Summit" www.un.org/geninfo/bp/enviro.html

Waste Reduction Technologies

1. http://www.epa.gov/reg5rcra/wptdiv/wastemin/

2. http://www.epa.gov/guide/sbf/final/dd/finalddpart2.pdf

3. http://www.dwmg.org/waste minimization through solid.htm

4. http://www.gidatec.co.uk/Cuttings.htm

5. http://www.glossary.oilfield.slb.com/Display.cfm?Term=weighted%20mud

6. http://www.aade.org/technicalpapers/2001conference/
Environmental/AADE%2012.pdf

7. http://www.aade.org/technicalpapers/2001conference/
Environmental/AADE%2013.pdf

Landfilling
1. Wright, Ross, and Tagawa, 1989

Landfarming
1. http://www.epa.gov/swerust1/cat/landfarm.htm
2. http://www.envirotools.org/remediation/remedisoilsed.shtml
3. http://www.nmenv.state.nm.us/ust/cl-landf.html
4. http://www.frtr.gov/matrix2/section4/4_13a.html
5. http://www.enviro.nfesc.navy.mil/erb/restoration/technologies/
remed/bio/bio-08lasp

Thermal Desorption
1. http://www.clu-in.org/products/citguide/thermdsp.htm
2. http://www.frtr.gov/matrix2/section4/4-26.html
3. http://www.cpeo.org/techtree/ttdescrip/thedesop.htm
4. http://www.envirotools.org/factsheets/fs_thermal_desorption.pdf
5. http://www.environmentalcenter.com/articles/article1218/article1218.htm
6. http://www.epa.gov/oust;%20questions/pubs/tum_ch6.pdf
7. http://www.epa.gov/tio/tsp/download/tdissue.pdf
8. http://www.epa.gov/swerust1/cat/lttd.htm
9. http://enviro.nfes.navy.mil/erb/restoration/technologies/remed/
phys_chem./phC-36.asp

Incineration
1. http://www.environmental.usace.army.mil/library/pubs/tsdf/sec3-1/sec3-1.html
2. http://www.enviro.nfesc.navy.mil/erb/restoration/technologies/remed/
phys_chem/phc-17.asp
3. http://www.frtr.gov/matrix2/section4/4-23.html

Bioremediation

1. http://www.cramont.it/projects/bioremediation.htm
2. http://www.cplpress.com/contents/C19 1.htm
3. http://www.isc.tamu.edu/PICS/FY95-proposal-part2/subsubsection3_14 2.html
4. Flathman, Paul E., Jerger, Douglas E., Exner, Jurgen H., Bioremediation: Field Experience (CRC Press, 1994)
5. King, R. Barry, Long, Gilbert M., Sheldon, John K., Practical Environmental Bioremediation (Lewis Publishers, 1992)
6. Wagenet, R. J., Bouma, J., The Role of Soil Science in Interdisciplinary Research, SSSA Special Publication Number 45 (1996)
7. Quoted in the U.S. Department of Energy Natural and Accelerated Bioremediation Research Program (NABIR), DOE Office of Health and Environmental Research Home Page, Internet, January 1996.
8. Strumm and Morgan (1981) and Metcalf and Eddy (1991)

Adsorption, Ion Exchange, Air Stripping

1. Parkhurst, J. D., F. D. Dryden, G. N. McDermott, and J. English, Pomona Activated Carbon Pilot Plant, J. Wat. Poll. Cont. Fed., 39:2,69–81 (1967).
2. Weber, W. J. Jr., C. B. Hopkins and R. Bloom, Physicochemical Treatment of Waste, J. Wat. Poll. Cont., 42:2, 83–99 (1970).
3. Gustafson, R. L., R. L. Albright, J. Heisler, J. A. Lirio, and A.0 Reid, Adsorption of Organic Species by High Surface Arc Styrene-Divinylbenzene Copolymers, I & EC Prod. Res. & Dev., 7:2, 107–115 (1968).
4. Kim, B. R., V. L. Snoeyink, and F. M. Saunders, Adsorption of Organic Compounds by Synthetic Resins, J. Wat. Poll. Cont. Fed., 48:1, 120–133 (1976).
5. Paleos, J., Adsorption from Aqueous and Nonaqueous Solutions on Hydrophobic and Hydrophilic High Surface-Area Copolymers, J. Coll. & Inter. Sci., 31:1, 7–18 (1969).
6. McGuire, M. J., and I. H. Suffet, Adsorption of Organics from Domestic Water Supplies, J. Am Water Works Assoc., 621–636 (1978).
7. Chudyk, W. A, V. L. Snoeyink, D. Beckmann, and T. J. Temperly, Activated Carbon versus Resin Adsorption of 2-Methylisoberneol and Chloroform, J. Am. Water Works ASSOC., 529–538 (1979).

8. Fox, C. R., Removing Toxic Organics from Waste Water, Chem. Eng. Prog., 70–77 (1979).

9. Kennedy, D. C., Treatment of Effluent from Manufacture of Chlorinated Pesticides with a Synthetic, Polymer Adsorbent, Amberlite XAD-4, Env. Sci. & Tech., 7:2, 138–141 (1973).

10. Van Vliet, B. M., and W. J. Weber, Comparative Performance of Synthetic Adsorbents and Activated Carbon for Specific Compound Removal from Waste, J. Wat. Poll. Cont. Fed., 53:11, 1585–1598 (1981).

11. Weber, W. J. Jr., Physicochemical Processes for Water Quality Control, Wiley Interscience, New York, NY. (1972).

12. Belfort, G., Adsorption on Carbon: Theoretical Considerations, Envi. Sci. & Tech., 14:8, 910–914 (1980).

13. Weber, W. J., and J. C. Morris, Equilibrium and Capacities for Adsorption on Carbon, J. San. Eng. Div. of ASCE (1964).

14. Giusti, D. M., R. A. Conway, and C. T. Lawson, Activated Carbon Adsorption of Petrochemicals, J. Wat. Poll Cont. Fed., 46:5, 947–965 (1974).

15. Bahrani, K. S., and R. J. Martin, Adsorption Studies using Gas-liquid Chromatography: Effect of Molecular Structure, Water Res., 10, 731–736 (1976).

16. Butler, J. A., and C. Ockrent, Studies in Electric Capillarity, J. Phys. Chem., 34, 2841–2859 (1930).

17. Schay, G. J., F. P. Fejes, and J. Szathmary, Studies on the Adsorption of Gas Mixtures—Statistical Theory of Physical Adsorption of the Langmuir-type in Multicomponent Systems, Acta Chem. Acad. Sci. Hungary, 12, 229 (1957).

18. Sweed, N. H., and R. A. Gregory, Parametric Pumping: Modeling Direct Thermal Separation of Sodium Chloride-Water in Open and Closed Systems, AIChE J., 171–176 (1971).

19. Crittenden, J. C., Mathematical Modeling of Fixed Bed Absorber Dynamics: Single Component and Multicomponent, Ph.D. Thesis, University of Michigan, Ann Arbor, MI. (1976).

20. Troxler, W. L., C. S. Parmele, and D. A. Barton, Survey of Industrial Applications of Aqueous-Phase Activated-Carbon Adsorption for Control of Pollutant Compounds from Manufacture of Organic Compounds, U.S. EPA, Office of Research and Development, Cincinnati, Ohio (1983).

21. Johnston, W. A., Designing Fixed-Bed Adsorption Columns, Chem. Eng., 87–92, November (1972).

22. Hutchins, R. A., New Method Simplifies Design of Activated-Carbon Systems, Chem. Eng., 133–138, August (1973).

23. Lukchis, G. M., Adsorption Systems Part 1: Design by Mass-Transfer-Zone Concept, Chem. Eng., 111–116, June (1973).

24. Weber, W. J. Jr., and J. C. Crittenden, MADAM I—A Numeric Method for Design of Adsorption Systems, J. Wat. Poll. Cont. Fed., 47:5, 924 (1975).

25. Grinstead, R. R., Copper-Selective Ion-Exchange Resin with improved Iron Rejection, J. of Metals, 31:3, 13–16 (1979).

26. Jones, C. K., and R. R. Grinstead, Properties and Hydrometallurgical Applications of Two New Chelating Ion Exchange Resins, Soc. of Chem. Ind., August 6 (1977).

27. Kennedy, C. D., Predict Sorption of Metals on Ion-Exchange Resins, Chem. Eng., 106–118, June 16 (1980).

28. Helfferich, F., Ion Exchange, McGraw-Hill Book Comp., Inc., NY. (1962).

29. Bio-Rad Laboratories, Chelex 100 Chelating Ion Exchange Resin for Analyses, Removal or Recovery of Trace Metals, Product Information 2020 (1976).

30. Gupta, A. K., Chemical Industry Developments, Incorporating CP $ E, 10:12, 15–25, Dec. (1976).

31. Perry, R. H., and C. H. Chilton, Chemical Engineering Hand-book, McGraw-Hill Book Comp., New York (1973).

32. Sullivan, K. M., T. E. Johnson, and F. C. Lenzo, Pilot Testing & Design of a Modular High-Temperature Air Stripping System, Proceedings of the Industrial Wastes Symposia, 57th Annual WPCF Conference, New Orleans, Louisiana, 109–119, Sept. 30–Oct. 4 (1984).

33. Treybal, R. E., Mass Transfer Operations, McGraw-Hill Book Co., New York, 3rd ed. (1980).

34. Sherwood, T. K., R. L. Pigford, and C. R. Wilke, Mass Transfer, McGraw-Hill Book Co., New York (1975).

35. McCarty P. L., Removal of Organic Substances from Water by Air Stripping, In: Control of Organic Substances in Water and Waste (B. B. Berger editor), Office of Research and Development, U.S. Environmental Protection Agency Washington DC, EPA-600/8-83-011, April (1983).

36. Stenzel, M. H., and U. S. Gupta, Treatment of Contaminated Groundwater with Granular Activated Carbon and Air Stripping, J. Air Poll. Cont. Ass., 35:12, 1304–1309 (1985).

37. Hodgman, C. D., R. C. Weast, and S. M. Selby (eds.), Handbook of Chemistry and Physics, 39th ed., Chemical Rubber Pub. Co., Cleveland, OH (1958).

38. Kavanaugh, M. C., and R. R. Trussel, Design of Aeration Towers to Strip Volatile Contaminants from Drinking Water, J. Wat. Works Ass., 72:12, 684–692 (1980).

39. Kincannon, D. F., and E. L. Stover, Treatment of Ground Water, In: Ground Water Pollution Control (By L. W. Canter and R. C. Knox), Lewis Pub. Inc., Chelsea, Michigan (1985).

40. Metcalf & Eddy, Inc., Waste Engineering: Collection, Treatment, Disposal, McGraw-Hill Book Comp., New York (1972).

41. Clark, J. W., W. Viessman, and Mark J. Hammer, Water Supply and Pollution Control, 3rd ed., Harper & Row Publishers, New York (1977).

42. Cawley, W., ed., Treatability Manual, Vol. III, Technologies for Control/ Removal of Pollutants, USEPA, 600-8-80-042-c (July, 1980).

Deep Well Re-injection

1. United States Environmental Protection Agency. Class II Injection Wells and Your Drinking Water. Office of Water. July 1994. EPA 813-F-94-003

2. http://www.frtr.gov/matrix2/section4-54.html

3. http://www.epa.gov/seahome/inject/src/title.htm

4. Bouffard, Karen. Hazardous waste wells likely to get EPA's OK. The Detroit News. May 15, 2003.

5. Domino, Andrew. Unsafe at Any Depth—Romulus Fights Toxic Well. From the Ground Up Ecology Center Newsletter. December 1999/January 2000.

6. http://www.ecocenter.org/199912/romulus.shtml

7. Harding, Dr. A Clean Look at Deep Well Isolation. Pollution Engineering. September 2003.

8. United States Environmental Protection Agency, Region 5. Land Disposal Restrictions Exemption Proposed. December 2002.

9. http://www.epa.gov/region5/water/uic/pubpdf/factsheet.pdf

10. Warikoo, Nirja. U.S. allows hazardous waste disposal in wells. Detroit Free Press. March 17, 2004.

ABOUT THE AUTHOR

Dr. Aloysius A Aguwa is the president of Altech Environmental Services Inc. (ALTECH), based in the United States, and the president of Chemical & Environmental Engineering Ltd. (CEEL), with headquarters at Owerri, Imo State, Nigeria. He is a recognized expert in pollution control, specializing in hazardous waste management. In addition to having published several articles, he has contributed chapters to various books. Both ALTECH and CEEL specialize in hazardous- and non-hazardous-waste management.

Dr. Aguwa has provided consulting services to industries, government agencies, and other institutions in the United States, Mexico, Canada, and Nigeria. Most of his work has dealt with the management of hazardous and non-hazardous waste through landfill operations. This includes landfill design, construction, and operation and maintenance. He has participated in providing remedial action at more than fifty landfills to bring them to compliance, and routinely provides technical support during environmental litigation. He has also been instrumental in the implementation of several innovative remedial technologies.

Dr. Aguwa's Ph.D. in environmental engineering was earned at the Illinois Institute of Chicago, following his B.S. and M.S. from Howard University in Washington DC. He belongs to several professional organizations, and is a diplomat of the American Board of Forensic Engineering and Technology. Dr. Aguwa is a recipient of the Chrysler Corporation's Chairman's Award. The Chairman's Award recognized Dr. Aguwa's outstanding achievement regarding the recycling of waste materials into bricks.

ABOUT THIS BOOK

Waste Management in the Oil Industry deals with strategies that are required to manage wastes emanating from oil exploration, development, refining, storage, and distribution in an environmentally friendly manner. Strategies identified range from waste-minimization programs to treatment options. The book is a useful tool for those responsible for implementing environmental activities in the oil industry.

978-0-595-41109-2
0-595-41109-6